Lecture Notes in Physics
Monographs

Springer
Berlin
Heidelberg
New York
Barcelona
Hong Kong
London
Milan
Paris
Singapore
Tokyo

The Editorial Policy for Monographs

The series Lecture Notes in Physics reports new developments in physical research and teaching - quickly, informally, and at a high level. The type of material considered for publication in the monograph Series includes monographs presenting original research or new angles in a classical field. The timeliness of a manuscript is more important than its form, which may be preliminary or tentative. Manuscripts should be reasonably self-contained. They will often present not only results of the author(s) but also related work by other people and will provide sufficient motivation, examples, and applications.
The manuscripts or a detailed description thereof should be submitted either to one of the series editors or to the managing editor. The proposal is then carefully refereed. A final decision concerning publication can often only be made on the basis of the complete manuscript, but otherwise the editors will try to make a preliminary decision as definite as they can on the basis of the available information.
Manuscripts should be no less than 100 and preferably no more than 400 pages in length. Final manuscripts should be in English. They should include a table of contents and an informative introduction accessible also to readers not particularly familiar with the topic treated. Authors are free to use the material in other publications. However, if extensive use is made elsewhere, the publisher should be informed. Authors receive jointly 30 complimentary copies of their book. They are entitled to purchase further copies of their book at a reduced rate. No reprints of individual contributions can be supplied. No royalty is paid on Lecture Notes in Physics volumes. Commitment to publish is made by letter of interest rather than by signing a formal contract. Springer-Verlag secures the copyright for each volume.

The Production Process

The books are hardbound, and quality paper appropriate to the needs of the author(s) is used. Publication time is about ten weeks. More than twenty years of experience guarantee authors the best possible service. To reach the goal of rapid publication at a low price the technique of photographic reproduction from a camera-ready manuscript was chosen. This process shifts the main responsibility for the technical quality considerably from the publisher to the author. We therefore urge all authors to observe very carefully our guidelines for the preparation of camera-ready manuscripts, which we will supply on request. This applies especially to the quality of figures and halftones submitted for publication. Figures should be submitted as originals or glossy prints, as very often Xerox copies are not suitable for reproduction. For the same reason, any writing within figures should not be smaller than 2.5 mm. It might be useful to look at some of the volumes already published or, especially if some atypical text is planned, to write to the Physics Editorial Department of Springer-Verlag direct. This avoids mistakes and time-consuming correspondence during the production period.
As a special service, we offer free of charge LaTeX and TeX macro packages to format the text according to Springer-Verlag's quality requirements. We strongly recommend authors to make use of this offer, as the result will be a book of considerably improved technical quality.
Manuscripts not meeting the technical standard of the series will have to be returned for improvement.
For further information please contact Springer-Verlag, Physics Editorial Department II, Tiergartenstrasse 17, D-69121 Heidelberg, Germany.

Giampaolo Cicogna Giuseppe Gaeta

Symmetry and Perturbation Theory in Nonlinear Dynamics

Springer

Authors

Giampaolo Cicogna
Dipartimento di Fisica, Università di Pisa
Via Buonarroti 2, Ed. B
56127 Pisa, Italy

Giuseppe Gaeta
Dipartimento di Fisica, Università di Roma
p. le A. Moro
00185 Roma, Italy
Present address:
Depto de Matematicas, Universidad Carlos III de Madrid
28911 Leganes, Spain

Library of Congress Cataloging-in-Publication Data.

Die Deutsche Bibliothek - CIP-Einheitsaufnahme

Cicogna, Giampaolo:
Symmetry and perturbation theory in nonlinear dynamics / Giampaolo Cicogna ; Giuseppe Gaeta. - Berlin ; Heidelberg ; New York ; Barcelona ; Hong Kong ; London ; Milan ; Paris ; Singapore ; Tokyo : Springer, 1999
(Lecture notes in physics : N.s. M, Monographs ; Vol. 57)
ISBN 3-540-65904-8

ISSN 0940-7677 (Lecture Notes in Physics. Monographs)
ISBN 3-540-65904-8 Springer-Verlag Berlin Heidelberg New York

Typesetting: Camera-ready by the authors
Cover design: *design & production*, Heidelberg

Printed on acid-free paper
SPIN: 10644440 55/3144/du - 5 4 3 2 1 0

Preface

Symmetry has been a major ingredient in the development of quantum perturbation theory, and it is a basic ingredient of the theory of integrable (Hamiltonian and non-Hamiltonian) systems; yet, the use of symmetry in the context of general perturbation theory is rather recent. From the point of view of nonlinear dynamics, the use of symmetry has become widespread only through equivariant bifurcation theory; even in this case, attention has been mostly confined to linear symmetries.

Also, in recent years the theory and practice of symmetry methods for differential equations has become increasingly popular and has been applied to a variety of problems (following the appearance of the book by Olver [262]). This theory is deeply geometrical and deals with symmetries of general nature (provided that they are described by vector fields), i.e. in this context there is no reason to limit attention to linear symmetries.

In this book we look at the basic tools of perturbation theory, i.e. *normal forms* (first introduced by Poincaré about a century ago) and study their interaction with symmetries, with no limitation to linear ones. We focus on the most basic setting, i.e. systems having a fixed point (at the origin) and perturbative expansions around this; thus our theory is entirely local.

We have tried to give a reasonably self-contained discussion, so the first three chapters deal with symmetry, differential equations, and in particular dynamical systems in a quite general way (and with many exercises to help the beginner). This part represents a compact introduction to symmetric dynamical systems and does not contain new results.

In Chaps. IV and V we discuss normal forms in the presence of symmetries, both in the general (dynamical systems) case and in the Hamiltonian one, at the formal level, i.e. without considering the convergence of the power series entering in the theory. This convergence is studied in Chap. VI, again focusing on the role of symmetries. Many of the results presented in this part are quite recent, having been obtained in the last five years (mostly between 1994 and 1997), and some of them are actually completely new.

In the final part of the book we discuss some related problems. In Chap. VII we discuss the relation between symmetries of a dynamical system and its invariant manifolds, in particular the center manifold; in Chap. VIII we discuss an extension of normal forms theory, i.e. the "further normalization" of normal forms; and finally in Chap. IX we discuss a formal approach to asymptotic symmetries.

Obviously this is not a textbook, but we have tried to provide examples and (except in the final chapters) exercises to help the student and/or the beginner in the field.

The discussion given here is rather abstract, although the reader will find some references to applications (of a mathematical and physical nature). On the other hand, in the introduction we have explained in a non-technical way the general context of the theory discussed and developed in the main body of the book, also giving an extensive list of references dealing with (mostly, very recent) applications of normal forms theory.

The introduction is written in a different style, more similar to that of a short review paper; we hope that this will be useful to the reader who does not already have a strong motivation to study normal forms, without being repulsive to the expert.

Our subject is rather strictly delimited – which allowed us to keep the volume within a reasonable length – and we will not touch upon closely related fields. Together with the length argument, we feel excused for this on the one hand by the premature stage of the developement of nonlinear symmetry methods in some of these cognate fields, and on the other by the fact that several good expositions of the available results exist and are mentioned in the introduction.

In particular, we should mention that normal forms are obviously relevant to bifurcation problems, but equivariant bifurcation theory is exhaustively discussed in the book by Golubitsky, Schaeffer and Stewart [169] (and in the nice short book by Iooss and Adelmeyer [190] and in the forthcoming book by Chossat and Lauterbach [80]) in the case of linear symmetries, and not yet so developed in the case of nonlinear symmetries. We hope this book can help such a development.

Acknowledgements

First of all we would like to thank Springer-Verlag and in particular Prof. W. Beiglböck for the patience shown with our several postponements of the final deadline for completing this book.

Our friends, colleagues and collaborators helped to shape our vision of the topic, and also of the present volume. We would like, in particular, to warmly thank Dario Bambusi and Sebastian Walcher for asking good questions, making useful remarks, and correcting errors.

We also learned a lot from all the participants to the two SPT conferences ("Symmetry and Perturbation Theory", held in Torino, 1996, and Rome, 1998). Our thanks to them for sharing their knowledge and interesting discussions, and to Dario Bambusi and Toni Degasperis for the organization of these venues.

G. Cicogna's research was funded over the years by the Physics Department of the University of Pisa and by the INFN (Istituto Nazionale di Fisica Nucleare) section of Pisa, and by GNFM-CNR (National Mathematical Physics Group of the Italian National Research Council).

G. Gaeta benefited from the hospitality of IHES at the time when most of the volume was written, and later on of the Theory Group of the Physics Department in Rome "La Sapienza", and of the Mathematics Department of the Universidad Carlos III in Madrid in the final phase of writing. He would like to thank these institutions and in particular J.P. Bourguignon, L. Michel, A. Degasperis, E. Franco, L.A. Ibort and M.A. Rodríguez for their hospitality and help, and P. Collet and S. van Strien for discussions and specific encouragement concerning normal forms and his work in the field.

The final revision was conducted on the computer of M.A. Rodríguez (Universidad Complutense de Madrid), monopolizing it for a certain time; we thank him and the Departamento de Física Teórica II for this most friendly help.

Pisa and Rome, September 1999 *G. Cicogna* and *G. Gaeta*

Contents

Introduction 1

1. Perturbation Theory 3
2. Other Tools to Study Nonlinear Systems and Symmetry 6
3. Why Nonlinear Symmetries? 8
4. Why Normal Forms? 10
5. Normal Forms and Higher-Level Perturbation Theory 14
6. Direct Applications of Normal Forms Theory in Physics 16

I. Symmetry and Differential Equations 19

1. Geometrical Setting 21
2. Invariance of Equations and of Solutions 24
3. Reduction and Solution of Ordinary Differential Equations 29
4. Reduction and Solution of Partial Differential Equations 33
5. On the Application of Symmetry Methods 38

Appendix. The Prolongation Formula 39

II. Dynamical Systems 41

1. Dynamical Systems and Flows 41
2. Singular Points and Invariant Manifolds 43
3. Conserved Quantities 47
4. Perturbative Expansion 48
5. Poincaré-Dulac Normal Forms 51
6. Birkhoff-Gustavson Normal Forms 56
7. Bifurcation Theory 61

Appendix. Lie Transforms 64

III. Symmetries of Dynamical Systems 67

1. Symmetries of Dynamical Systems 68
2. Lie-Point Time-Independent Symmetries 69
3. Constants of Motion and the Module Structure of the Symmetry Algebra 71
4. Symmetry and Topology of Trajectories 72
5. Time-Dependent Symmetries 75
6. Orbital Symmetries 77
7. Approximate Symmetries 78

Appendix. On the Module Structure 81

IV. Normal Forms and Symmetries for Dynamical Systems 83

1. Perturbative Expansion of Determining Equations 83
2. Recursive Determination of Symmetries 85
3. Approximate Symmetries 88
4. Symmetry Characterization of Poincaré-Dulac Normal Forms 89
5. Nonlinear Symmetries and Normal Forms 91
6. Linear Symmetries and Normal Forms 96
7. On Linear and Nonlinear Symmetries 97
8. Symmetry for Systems in Normal Form 99
9. Reduction to Normal Form of a Nilpotent Lie Algebra 101
10. Non-semisimple Normal Forms 104
11. The Linearization of a Dynamical System 108
12. Partial Joint Normal Form for Non-nilpotent Algebras 109

Appendix. Some Results on Matrices and Lie Algebras 112

V. Normal Forms and Symmetries for Hamiltonian Systems 115

1. Birkhoff-Gustavson Normal Forms 115
2. Birkhoff-Gustavson Normal Forms with Symmetries 118
3. The Case of d-Dimensional Algebras of Symmetries 124
4. Perturbative Construction of Symmetries 126
5. The Non-normal Case 128
6. The Normality of the Homological Operator 132

VI. Convergence of the Normalizing Transformations 135

1. Conditions Ensuring Convergence 135
2. Normalizing Transformations in the Presence of Symmetries 139
3. Convergence and Symmetries: A General Result 141
4. Convergence and Symmetries: A Special Case 144
5. Convergence in the Case of Hamiltonian Problems 148

VII. Invariant Manifolds 151

1. Some Preliminary Results on Flow Invariant Manifolds 151
2. Reduction to a Center Manifold 154
3. Normal Forms 155
4. Shoshitaishvili Theorem and Center Manifolds 156
5. Some Examples 158

VIII. Further Normalization 163

1. Higher-Order Terms in Poincaré Transformations 164
2. The Homological Operators 166
3. Non-uniqueness of Poincaré Normal Forms 166
4. Poincaré Renormalization 167
5. Renormalization by Iterated Normalizations 171
6. Examples: Planar Vector Fields 173
7. The Hamiltonian Case 177
8. Renormalized Forms in the Presence of Symmetry 179

IX. Asymptotic Symmetries 183

1. Notation and Basic Sets 184
2. Induced Actions on Functions and Equations 185
3. Symmetries and Asymptotic Symmetries 187
4. Asymptotic Symmetries and Space-Time Asymptotic Properties 189

References 193

Introduction

This volume deals with normal forms theory in the presence of (possibly nonlinear) symmetry. Normal forms theory, which was created by Poincaré in his study of nonlinear systems, is – together with averaging theory – a fundamental tool of perturbation theory, and as such it is of wide applicability to many, if not all, branches of Physics.

In the main body of the volume we will deal with the theory mainly from a rather abstract and general point of view, so in this introduction we want – as far as is possible within reasonable length – to put the theory into its context, discuss relations with other perturbative and non-perturbative approaches, and in particular discuss the physical motivation underlying the theory together with its physical applications. We will also give a number of (mostly very recent) references to help the reader look for applications to different fields of physics and mathematics.

In a way, this introduction is conceived for readers just casually looking at the book on the shelves of a library but is wondering if normal forms can help their research concerning a specific field of physics (or mathematics): readers who already have an interest in normal forms and/or symmetry methods should just proceed to Chap. I (or a later one according to their knowledge and interests).

Obviously, perturbation theory is relevant whenever we deal with perturbations of a simple (linear or integrable) system or with a system that can be seen as such. In this sense, it is difficult to find a branch of physics in which it is not relevant. Indeed, in most cases when we deal with all except the very simplest, isolated and idealized systems (harmonic oscillators, free fields, a solar system with only one planet, the hydrogen atom[1]) what we do is precisely to resort to a perturbative study of the system of interest, seeing it as a perturbation of a simple system (e.g. small amplitude dynamics for a nonlinear oscillator can be seen as a perturbation of linear oscillator dynamics), or as an assembly of interacting simple systems, where the interaction is seen as a perturbation (e.g. "real" solar systems with several planets, atoms – or nuclei – different from hydrogen, a molecule, a particle beam in an accelerator, or an artificial satellite, etc.).

Thus, what we present here is a tool of interest well beyond what has been our original source of interest and physical motivation – although the expert reader will understand that this was mainly connected with classical mechanics of point particles, which is also the historical source of perturbation theory – and we believe an abstract presentation will have the advantage of making the results

[1] It is not an accident that these examples correspond to *integrable systems*.

immediately available to people working in different fields of physics.[2] It should be stressed that nonlinear dynamics is of interest not only to theoreticians, but also to experimentalists (and engineers), as it is indicated, for example, by the sections devoted to it in leading journals.

In this sense, this work should not be seen as theoretical physics, or mechanics, but rather as *mathematical methods of physics* (i.e. mathematics which happen to be relevant to physics). We hope that readers with different backgrounds will find different ways of looking at it from their specific points of view; and that at least some of them will also find it useful in their work on concrete systems quite far from our experience and original motivation.

For the same reason, i.e. to make the work accessible to a wider audience – and in particular to students – we provide some introductory chapters with a simple discussion of dynamical systems and of symmetries; readers already familiar with these topics can safely skip the first chapters, and readers desiring a more complete discussion of the general foundations can refer to one of the excellent existing books.

On the other hand, an abstract presentation like the one we adopt in the main body of the volume could make it more difficult to settle our discussion in the right context, so we are now going to discuss this qualitatively and mention some of the physical (motivation and) applications of normal forms theory, in particular when symmetries are also present.

It is indeed an experimental fact that many systems of interest in physics have *symmetries.* In most – if not all – cases the symmetries are not just found to be an "accidental" property of the equations describing the systems, but actually are a fundamental property of the physical system. Thus, for a physicist it is entirely natural to devote special attention, in all cases and in all fields of mathematics, to the features which are brought into the theory by the presence of symmetries; what we do in this volume is indeed to study *normal forms theory in the presence of (possibly nonlinear) symmetries*, for dynamical systems.

It is interesting to note that the role of symmetries within perturbation theory for ordinary differential equations (ODEs) received attention quite late[3] [38–40,121], and anyway much later than perturbation theory for partial differential equations has. Actually, this is easily explained by the fact that the study of symmetric perturbation theory for PDEs was performed by physicists in the frame of quantum mechanics [339]. As remarked by Hermann, in Weyl's book of 1955 it was already made clear that *"most of the successful applications of quantum theory are based on general group-theoretic principles and ... perturbation theory"* [182, p.160]. It is remarkable that the role of symmetry in general pertur-

---

[2] And in nonlinear mathematics, of course. Actually, in recent years it has become increasingly clear that nonlinear dynamics is also relevant to engineering, chemistry, biology and, lately, also to finance and economics, which are all fields we are completely incompetent to discuss.

[3] The role of symmetries for Hamiltonian systems, also in connection with perturbation theory, was realized more promptly, but still emphasized only in relatively recent times.

bation theory was not really studied for the next quarter of a century or more[4], until, starting in the 1970s, it received interest in the context of *bifurcation theory* and pattern formation (see below). Even once studies of the interaction between symmetry and perturbation theory were begun, attention was for a long time, limited to *linear* symmetries only[5], in contrast to our approach here.

It should be mentioned that, of course, symmetries received diffuse attention in the context of *integrable systems* (both finite-dimensional and infinite-dimensional ones). We mention e.g. [2,13,17,132–136,200,230,237,239,261,262, 269,277,312].

1. Perturbation Theory

The general problem of *perturbation theory* can be described as follows: we have a system (say, a differential equation) S_0, on which we have determined information (e.g. the complete solution, or a qualitative property of it, as stability of a given solution); we then consider "nearby" systems S_ε, and would like to infer information on this from the information we have on S_0. Needless to say, first of all we should be able to define the concept of a "nearby system"; however, once we have done this, in general – i.e. for arbitrary systems and arbitrary properties of these – there is no reason to expect that S_ε behaves, for the characteristic we are interested in, as S_0.

An important part of this century's mathematics has been devoted to the definition and study of "structurally stable" systems, i.e. of systems whose qualitative behavior is stable under small perturbations, and to the understanding and classification of the qualitative changes ("bifurcations") that can occur when we deal with non-structurally stable systems.

The physical counterpart of these concepts is particularly direct in statistical physics and field theory, with the modern theory of critical phenomena, corresponding to universality classes and phase transitions. More generally, this mathematical approach is particularly appealing to physicists, as it focuses on qualitative generic behavior of systems and on transitions among different such behaviors which the system can have under different conditions (external control parameter, energy, temperature, etc.).

---

4 In the frame of the study of Hamiltonian systems, integrability in the Liouville-Poincaré-Arnold sense – which corresponds to the possibility of mapping the systems to a set of harmonic oscillators, which themselves are highly symmetric – is guaranteed and characterized by the presence of sufficiently many constants of motion, i.e. commuting symmetries, and even the symplectic structure can be seen as a kind of symmetry property. Diffuse attention was devoted to such systems and their perturbation a lot earlier than the date mentioned (just think of KAM theory), but it seems the symmetry properties have played little role in discussing perturbations.

5 These are of course the most commonly met, and thus the first we have to study, but by no means the only ones. Moreover, symmetries which are linear in one system of coordinates could be nonlinear in another, see the discussion below.

This line of mathematical research was started by Poincaré at the turn of the century; seminal developements in the next half century or so are usually ascribed to Andronov, Landau[6] and Pontrijagin in URSS, to Hopf and Kupka in Europe, and to Birkhoff, Morse, Palais and Smale in the USA. Developements became too numerous and fast in the last thirty years – also establishing links with other branches of mathematics, and with relevant contributions by other mathematical schools, e.g. the Brasilian, Dutch and French ones – to mention even only the main contributions.

This activity has been particularly important in the framework of the study of *dynamical systems* and is reflected, for example, in the volumes of Springer's "Encyclopaedia of Mathematical Sciences", devoted to dynamical systems. It should be stressed that such an interest is not only of a theoretical nature, but also of the foremost importance for concrete applications, where we can never obtain absolutely precise information on the equations governing a real system (as opposed to the idealized mathematical model of it that we study) and on initial conditions. Only predictions which are "robust" with respect to perturbation of the model and/or of initial conditions are relevant for practical applications.

It would be out of place to discuss this central concept of mathematics here, and we will just refer, for example, to [3,11,12,173,300]. In this volume we are only interested in finite-dimensional dynamical systems, i.e. in autonomous systems of first-order ordinary evolution equations,

$$\dot{x} = f(x) , \tag{1}$$

where x is a point on a smooth manifold M, and $f : M \to \mathrm{T}M$ is a function (section of the tangent bundle) defining a vector field on M. We will suppose, for ease of discussion, that we have already guaranteed the existence and unicity of solutions to (1) with Cauchy initial condition $x(t_0)$ and actually that f is analytic.

Suppose we know a solution $x(t) = a(t)$ for (1); we would then like to obtain information on nearby solutions $x(t) = a(t) + y(t)$ with $y(t)$, or at least $y(t_0)$, small in the natural norm induced by the metric on M.

[6] The reader could be surprised to see the name of Landau among those of distinguished mathematicians; we would like to quote, in his respect, a few phrases from chapter V of Arnold's monograph on catastrophe theory [14], to which we also refer the interested reader for interrelations between this and perturbation theory (the same volume also contains a deep review of bifurcation theory [3]): *"Among the physicists who systematically applied catastrophe theory before it was born, one should particularly single out L.D. Landau. In his hands, the art of throwing away inessential terms of Taylor series, retaining the higher order, but "physically important", terms yielded many excellent results, now included within catastrophe theory. Thus, for example, in work done in 1943 on the generation of turbulence, by this method Landau directly wrote out the equation of the Hopf bifurcation ... . Landau's theory of phase transition of the second kind reduces to the analysis of bifurcations of critical points of symmetric functions".*

The simplest such case is that where the known solution is just $x(t) = x_0$ [i.e. x_0 is a fixed point for (1)] and we are interested in the behavior of solutions which start near x_0; we stress that in general these solutions will not remain near x_0, but we can be interested in knowing their behavior at least for the time in which they stay within a radius r from x_0, or to know how long will they stay within such a radius. It is quite clear that such a problem is completely *local*, and the geometry of M does not play any role. We can just consider a local chart including x_0 and local coordinates in it[7], so that we can equally well consider the problem in $\mathbf{R}^n$ rather than in M.

We can go on to consider $y = (x - x_0)$ and the linearization of the resulting equation, which we write as

$$\dot{y} \;=\; g(y) \;:=\; f(y + x_0) \tag{2}$$

around 0 [that is, of (1) around x_0], obtaining the *linearized equation*

$$\dot{y} \;=\; Ay \tag{3}$$

with $A = (Df)(x_0) = (Dg)(0)$. We are able to study (3) in some quite detail, but what is the relation between the solutions (say, with same initial data near x_0) to (1) and to (3) ?

This can be seen in a slightly different – but equivalent – way: if we rescale coordinates as $\xi = \varepsilon y$ (e.g. so that $|y| = r$ corresponds to $|\xi| = 1$), the equation (2) reads

$$\dot{\xi} \;=\; F(\xi, \varepsilon) \tag{4}$$

with $F(0, \varepsilon) = 0$ for all ε. When we expand F (assume it is analytic around 0, which is obviously true if f is analytic in M) in Taylor series, and call F_k the term homogeneous of degree k in the ξ, (4) reads

$$\varepsilon\dot{\xi} \;=\; \sum_{k=1}^{\infty} \varepsilon^k F_k(\xi) \ ,$$

and upon dividing both sides by $\varepsilon \neq 0$ we obtain

$$\dot{\xi} \;=\; A\xi + \sum_{k=2}^{\infty} \varepsilon^{k-1} F_k(\xi) \ , \tag{5}$$

where A is the same matrix as before. This can be seen as a linear equation, like (3), for $\varepsilon = 0$, plus a small perturbation of $O(\varepsilon)$ [more generally, of $O(\varepsilon^s)$ where $k_0 = s + 1$ is the order of the first F_k which is not identically zero].

Again, we can obtain quite complete results for the linearized equation in which we set $\varepsilon = 0$, but how many of these will actually apply to the full equation ?

This kind of problem is, of course, appearing in all branches of physics, and rests on the basic fact that we are able to analyze linear equations much better

[7] Unless the application we are interested in forces us to consider a neighbourhood r of radius larger than the one allowed to local charts around x_0.

than nonlinear ones, so that we try to consider linear approximations of all kinds of phenomena[8], trying to grasp some understanding of the full physical (nonlinear) picture.

We have discussed only the case of systems described by equations like (1) in which we know of a stationary solution, but the picture obviously applies to other kinds of problems via simple manipulations.

(*i*) First of all, for systems described by higher-order ODEs (e.g. second-order equations), we can obviously always rewrite the equations in the form of a first-order system.

(*ii*) For cases where we know of a specific – and physically interesting – solution $a(t)$ which is non-stationary, we can always go on to consider $z(t) = x(t) - a(t)$, so that the known solution becomes a fixed point for the equation describing the evolution of $z(t)$. Notice that in this case we would get a system

$$\dot{z} = f(z,t) , \tag{6}$$

which is *not* autonomous; however, by a standard technique associated with the name of Floquet (equivalently, studying the *monodromy operator* [12]), this can be mapped to an autonomous system. This kind of situation is of particular physical interest when $a(t)$ is a periodic or quasi-periodic solution, in which case we get concrete results [12]. Notice that in this case the problem can be influenced by the global geometry of M, as now we are actually considering a tubular neighbourhood of the reference solution $a(t)$.

(*iii*) Finally, there are cases in which we have, since the beginning, a system that can be interpreted as a completely solvable system (in the framework of dynamical systems, an integrable one), plus a small perturbation of order ε. Again, focusing on a specific solution of the unperturbed system, we arrive at a situation of the kind considered above. As above, if the specific solution under consideration is a fixed point, we are dealing with a local problem, but if it happens to be a periodic or multi-periodic one, the global geometry of the manifold can enter into play [13,17,207].

2. Other Tools to Study Nonlinear Systems and Symmetry

Perturbation theory is the most ancient tool for the study of nonlinear systems, having been used already by Laplace in his investigation of celestial mechanics, but of course it is not the only tool we can use. Among the other tools, those based on *topology* and on *symmetry theory* play a special role, in that they give general properties, valid independently of the nature – linear or nonlinear – of

[8] Needless to say, if we deal with infinite-dimensional physics, i.e. with fields, linear equations can still be too hard to solve, as quantum mechanics beyond the simplest systems clearly shows.

the system. The information we get by topological methods[9] is qualitative or partial – e.g. we know the value of a global invariant – but it is often precious in many respects. The information provided by symmetry considerations is maybe less general than the topological one (first of all, it applies only to symmetric systems!) but it has the advantage of being of a somehow more quantitative nature.

In this volume, we will not consider topological issues – which are more and more relevant in many branches of physics – and focus instead on the *interrelations between perturbation theory and symmetry.* Indeed, if we deal with symmetric systems, the presence of symmetry changes the genericity classes of the system[10], and it is natural that taking into account the symmetries of the system can be exceedingly useful in its study.

This is even more relevant if the mathematical systems we are studying actually arise from physical problems: not only do we know that many physical problems display a symmetry, but in many cases the symmetry is actually one of the defining properties of the system (think, for example, of gauge theories or the already mentioned Landau theory of phase transitions and the modern theory of critical phenomena, where the required invariance concerns the renormalization group action). Further, some fundamental symmetries are building blocks of the theories at the foundations of physics. In classical mechanics, one assumes homogeneity and isotropy of space, i.e. covariance under the Galilei group. This is changed to the Poincaré group in special relativity and to the full diffeomorphism group in general relativity. On a different level, the basic property of linearity in quantum mechanics can be interpreted as invariance of the wave equation under the dilation group[11] acting on the wave function ψ.

Consideration of symmetry has been a major tool in recent developements of local bifurcation theory initiated by Sattinger [287–289] and Golubitsky and Stewart [169] (see also [236], the works by Field and Richardson [128–131], the survey [126] and the books [80,190,191]) following the fundamental observation by Ruelle [278] that degenerate systems[12] in which the degeneration is entirely due to symmetry are, after all, non-degenerate modulo the symmetry and can be treated in a completely satisfactory way. Moreover, such a symmetry degen-

[9] It should be recalled that topology in its modern form is considered to start with the *Analysis Situ* of Poincaré [270] and so has been connected since the beginning with the study of nonlinear system and perturbation theory.

[10] In this respect, we would like to quote the best explication we know [308, p. 235] of the differences between the symmetric and the non-symmetric cases: *"Symmetries complicate things and simplify them. They complicate them by introducing exceptional types of behavior, increasing the number of variables involved, and making vanish things that usually do not vanish. They simplify them by introducing exceptional types of behavior, increasing the number of variables involved, and making vanish things that usually do not vanish. They violate all the hypotheses of our favourite theorems, yet lead to natural generalizations of these theorems".*

[11] This is obviously equivalent to having a linear equation.

[12] Here the degeneration alluded to is the one relevant for bifurcations, i.e. among critical eigenvalues.

eration is not removable, i.e. is again a structurally stable property and thus "really interesting".

The reader familiar with bifurcations will be quite clear about the relation between the results discussed in this volume – and in general of normal form theory – with (symmetric) bifurcation theory; however, discussing this is out of our present scope, and would take us too far away. We will be satisfied with referring to [190] and to [164,319] for an excellent discussion of such interrelations without undue technical hardship; for a short discussion of the basics of symmetric bifurcation theory we refer to [308], and we refer to [143] for a discussion aimed at physicists, and [105] for one specifically suited to fluid dynamics; see also [85]. For an updated overview of the interrelations between bifurcation and symmetry, see [169] and the forthcoming book by Chossat and Lauterbach [80].

3. Why Nonlinear Symmetries?

In most cases of physical interest, the symmetries one has to consider are actually *linear* ones (e.g., rotations), with the notable exception of *scaling* symmetries, which, however, have linear infinitesimal Lie generators. Also, in many cases we are interested – in particular for finite-dimensional systems – in the *compact* symmetry group, and a well-known theorem by Mostow [249,268] guarantees that for such groups any finite-dimensional representation can be embedded in a linear representation (in general, of dimension higher than the original one).

So, is our study of nonlinear symmetries only due to a desire for completeness and generality? Not at all, and this is for several reasons.

First of all, nonlinear symmetries can occur in physically relevant systems (see e.g. [262,55]). On the one hand they do not necessarily generate a compact group[13], and on the other, even when we have compact groups, for which in principle (but see the remarks below) we could embed the system in higher dimension, getting a linear group action, learning how to deal with nonlinear actions can be relevant in view of extension to a infinite-dimensional situation. Moreover, the "higher dimension" in which we have to embed our system to get linearization of the group action can be exceedingly high, making the task computationally unviable.

But there is also a stronger point, related to the fact that Mostow theorem is not constructive: we know such a linearization of group action exists, but have no clue, from the general theorem, about how to realize it. It can well happen that in order to determine the transformation enabling us to linearize the group action, we first have to completely solve our system. Just think of the case in which we have a harmonic oscillator $\ddot{x} = -x$ expressed in "wrong coordinates"

[13]Obvious examples of noncompact groups are the translation and the scaling groups; scaling can for example be combined with rotation to have "spiralling" symmetries, conjectured to be relevant in galaxy formation.

(e.g. chosen by a particularly unfriendly colleague[14]), i.e. in coordinates $y = \Phi(x)$ obtained via a highly nonlinear transformation Φ: clearly the symmetries of the harmonic oscillator [scaling, and rotation in the $(x, \dot{x})$ plane, plus reflections] are still present in the y coordinates, but expressed through a highly nonlinear action.

If we first attack the problem by using the y coordinates – which is exactly what happens when we study a nontrivial integrable system before discovering the suitable change of coordinates to explicitely integrate it – we can be aware of the nonlinear symmetries, but expressing them in linear form requires that we move to the x coordinates, i.e. to integrate the system. Obviously this is not satisfactory: we would like to use the information we can get on symmetries in order to help in the integration of the system, not the other way round !

Let us now consider two simple examples of systems with nonlinear symmetries.

Example 1. We will consider a system defined in R^2 with coordinates $\mathbf{x} \equiv (x, y)$, given by

$$\begin{aligned} \dot{x} &= -y + x^2 - y^2 , \\ \dot{y} &= x + 2xy . \end{aligned}$$

This has the nonlinear symmetry given by the vector field

$$(x + 2xy)\frac{\partial}{\partial x} + (y - x^2 + y^2)\frac{\partial}{\partial y} .$$

The system is linearized by the transformation

$$x \to x' = \frac{x}{x^2 + (y+1)^2} \quad , \quad y \to y' = \frac{y + x^2 + y^2}{x^2 + (y+1)^2} ,$$

under which the above symmetry is nothing other than a dilation:

$$x'\frac{\partial}{\partial x'} + y'\frac{\partial}{\partial y'} .$$

Note that this system is not hyperbolic, so we could not even apply the Hartmann-Grobman theorem to guarantee the topological equivalence with its linear part.

Example 2. We consider again a system in R^2 with coordinates $\mathbf{x} \equiv (x, y)$, this time given by

$$\begin{aligned} \dot{x} &= x + y - x(x^2 + y^2) , \\ \dot{y} &= -x + y - y(x^2 + y^2) . \end{aligned}$$

This admits as symmetry the vector field

[14] We would like to thank G. Marmo – who is always very friendly and would never choose wrong coordinates for us – for this specially suggestive way of formulating the problem.

$$\left(x - x(x^2+y^2)\right)\frac{\partial}{\partial x} + \left(y - y(x^2+y^2)\right)\frac{\partial}{\partial y}$$

(equivalent to the superposition of the flow itself and a rotation). The system is linearized by

$$x \to x' = \frac{xr}{\sqrt{1-r^2}} \quad , \quad y \to y' = \frac{yr}{\sqrt{1-r^2}} \qquad (r^2 = x^2 + y^2) \ ,$$

and the symmetry is again just a dilation in the x', y' coordinates. We have not yet mentioned the problem of convergence of the normalizing transformation, but we note – for the reader who knows it, or for later use – that the linearizing transformation is convergent in the open disk $|\mathbf{x}| < 1$.

There is, of course, also a mathematical motivation to study nonlinear symmetries: the modern geometrical theory of symmetries of differential equations [4,32,55,146,262,263,305] – which is highly developed and successfully applied to a great variety of physical problems – does not seem to assign any special role to linear symmetries, so it is natural, even from an "interior" point of view of mathematical theory, to investigate the extension of the known results for linear symmetries to general, i.e. nonlinear, ones.

At the other extreme – in a way – of the spectrum of motivation, if in concrete applications one finds a symmetry which acts nonlinearly, it is quite desirable to be able to use it directly, i.e. without necessarily passing to coordinates in which it acts linearly – even if this is possible a priori and one is, moreover, actually able to do it – both in order to avoid extra work and because other kinds of considerations could require us to use different coordinates: thus it is surely better to be able to deal with symmetries in whatever coordinates we are using.

We expect that readers interested in different aspects of perturbation theory (or merely in its concrete applications) could find one or the other of these motivations more convincing; we will summarize all of them by saying (as trivial as it is) that being able to deal with nonlinear symmetries is an advantage over being able only to deal with linear ones, whatever the problem we are interested in.

4. Why Normal Forms?

In this volume we will study normal forms, and for completeness we also provide basic definitions and results, so that it is a self-contained investigation of normal forms from the point of view of symmetry. Readers already familiar with the application of normal forms will (hopefully) be attracted by the possibility of combining them with symmetry, to make them an even stronger tool and improve their utility, which they already know. These readers can safely skip this section.

But the reader who is not familiar with normal forms (e.g. the reader who is familiar with symmetry and sees here a way of using it in nonlinear dynamics, or the student who has not been previously exposed to a course on perturbation theory) can after all wonder what the practical use of this theory is. In the main body of the book we will discuss some of the uses of normal forms for the study of dynamical systems, mainly the qualitative one (classification of systems and of their qualitative behavior), and possibly for finding explicit solutions; for these the reader is referred to later chapters. In this section we would just like to illustrate the reason for introducing normal forms from a point of view not discussed later on but also natural to physicists, i.e. for concrete (possibly but not necessarily approximate) computations, deferring the discussion of their usefulness for other purposes.[15]

Let us consider a dynamical system in $\mathbf{R}^n$,

$$\dot{x} \;=\; f(x) + \varepsilon g(x;\varepsilon) \;, \tag{7}$$

with both f and g analytic functions of their argument.[16] We denote the solutions of the system (7) which satisfy $x(0) = x_0$ by $\Phi_\varepsilon(x_0, t)$ and suppose we know about the behavior of this system for $\varepsilon = 0$, i.e. about $\Phi_0(x_0, t)$.

We would then like to have information about the behavior of the full system, and in particular we would like to know if, for t small enough, the distance

$$\rho_\varepsilon(t) = |\, \Phi_\varepsilon(x_0, t) - \Phi_0(x_0, t)\, | \tag{8}$$

remains small.

Suppose that in a ball $\mathcal{B}$ centered in x_0, $f(x)$ admits a Lipshitz constant C_f and $|g(x;\varepsilon)|$ is bounded by a constant $M_g(\varepsilon)$: in this case we know that while both $x(t) = \Phi_\varepsilon(x_0, t)$ and $\xi(t) = \Phi_0(x_0, t)$ remains in $\mathcal{B}$,

$$\begin{aligned} \frac{\mathrm{d}\,\rho_\varepsilon(t)}{\mathrm{d}t} &\leq |\dot{x}(t) - \dot{\xi}(t)| \;=\; |f(x(t)) - f(\xi(t)) - \varepsilon g(\xi(t), \varepsilon)| \\ &\leq |f(x(t)) - f(\xi(t))| + |\varepsilon g(\xi(t), \varepsilon)| \;\leq\; C_f\, \rho_\varepsilon(t) + |\varepsilon| M_g(\varepsilon) \;. \end{aligned} \tag{9}$$

Thus the positive quantity $\rho_\varepsilon(t)$, which has to satisfy $\rho_\varepsilon(0) = 0$, grows more slowly than the solution $r(t)$ to

$$\frac{\mathrm{d}r}{\mathrm{d}t} \;=\; a\,r \;+\; b \;, \tag{10}$$

where $a = C_f$, $b = M_g|\varepsilon|$. As the solution to (10) satisfying $r(0) = 0$ is

$$r(t) \;=\; (b/a)\;\left[e^{at} - 1\right] \;, \tag{11}$$

we conclude that while both $x(t)$ and $\xi(t)$ are in $\mathcal{B}$,

$$\rho_\varepsilon(t) \;\leq\; |\varepsilon|\; \frac{M_g}{C_f} (e^{C_f t} - 1) \;. \tag{12}$$

[15]For this we just refer to [3,12], for the time being.

[16]For example, (7) could be (5) considered above, which is the case we are really interested in.

Note that if $x(t) \in \mathcal{B}$ is at a distance greater than $\rho_\varepsilon(t)$ from the border of $\mathcal{B}$, we are automatically guaranteed $\xi(t) \in \mathcal{B}$, so we can guarantee (12) without solving for the full system $\varepsilon \neq 0$.

Thus, if we are interested in having the solution to (7) within an error δ, we can guarantee that the solution to the system with $\varepsilon = 0$ is within this error from the true solution only for a time[17]

$$t_\varepsilon \; < \; \frac{1}{C_f} \, \ln \left[1 + \frac{\delta C_f}{|\varepsilon| M_g} \right] \; . \tag{13}$$

Suppose now that in (7) we had

$$\dot{x} \; = \; f(x) \; + \; \varepsilon^k \, g(x; \varepsilon) \; ; \tag{7'}$$

it is clear that by repeating the same discussion we would get

$$\rho_\varepsilon(t) \; \leq \; |\varepsilon|^k \, \frac{M_g}{C_f} (e^{C_f t} - 1) \; , \tag{12'}$$

$$t_\varepsilon \; < \; \frac{1}{C_f} \, \ln \left[1 + \frac{\delta C_f}{|\varepsilon|^k M_g} \right] \tag{13'}$$

and in particular for $\delta = \varepsilon > 0$ we get

$$\lim_{\varepsilon \to 0+} t_\varepsilon \; = \; M_g^{-1} \, \varepsilon^{-(k-1)} \; . \tag{14'}$$

Thus if we were able to find a change of coordinates under which (7) is changed into (7′) with $k > 1$, and still be able to solve the problem for $\varepsilon = 0$ (note that f and g, and thus C_f and M_g, are changed when we change coordinates, so that there are several subtleties, which we will not discuss at all, on the actual gain of control), without increasing too much the constant M_g, we would be able to get a better approximation for the solutions of the full system in terms of the solution to the simplified system corresponding to $\varepsilon = 0$. This "better approximation" roughly means, in terms of the notation employed above, that $\rho_\varepsilon(t)$ would be smaller at the same t, and that it would remain within the prescribed error for a longer time.

One of the things we *can* actually do with normal form theory, and the problem formally solved by Poincaré in his thesis, is precisely find a constructive way to build these changes of coordinates.[18] In particular, if we start from a generic system of the form (5), i.e.

$$\dot{\xi} \; = \; A\xi + \sum_{k=2}^{\infty} \varepsilon^{k-1} F_k(\xi) \; ,$$

[17] For ease of discussion we assume $x(t), \xi(t) \in \mathcal{B}$.

[18] These will be of the form $x = y + h(y)$, with h a completely nonlinear (vector) function. Among other things, this guarantees the linear part of the equation is not changed at all.

we can with p Poincaré transformations formally change the system into

$$\dot{\xi} = A\xi + \sum_{k=p+2}^{\infty} \varepsilon^{k-1} G_k(\xi) , \tag{5'}$$

where the linear part remains the same, while the nonlinear ones are changed and start not with quadratic terms (of order ε), but with terms of order $p+2$ (of order ε^{p+1}).

Here, "generic" means that the matrix A has to satisfy a (nonresonance) condition which is generally satisfied. As for the "formally", this alludes to the fact that the required change of coordinates is constructed as a series whose convergence is not guaranteed. This is related to the fact that one should avoid M_g growing "too much" (i.e. more than of a factor $1/\varepsilon$, or diverges) in the transformation pushing the perturbation one order higher in ε.

These concepts and problems will obviously be discussed in detail in the main body of the book; here we just want to stress what the computational use of normal forms theory is, to provide motivation to the reader.

We would just like to add, at this stage, that even in the cases where the full normalizing transformation does not converge, i.e. when we cannot take the limit of the above procedure for $p \to \infty$, one can still have a substantial advantage in performing a partial normalization, i.e. performing a finite number p of Poincaré transformations.

This is related to the fact that the series appearing in the theory are in general *asymptotic* ones, so that there is an optimal number p of normalization steps. However, we will not discuss this aspect of the theory, and we will limit ourselves to the two extreme cases: first (in Chaps. I–V) we discuss the formal theory, and later on we discuss conditions that ensure the normalizing transformation is actually convergent in some neighbourhood (small, but of full measure) of the origin.

The above discussion should show quite clearly that – contrary to what is often claimed – normal forms are not a way to "simplify the equation", at least if simplification is intended in the naive (and natural) meaning. If we have a system like, say,

$$\begin{aligned} \dot{x} &= -y \\ \dot{y} &= x - xy \end{aligned} \tag{15}$$

and start the normalization procedure, with the first step we eliminate the quadratic term xy, but at the same time generate an infinite number of higher-order terms.[19] Thus, it is hard to say that the normalized equation is simpler than the original system; on the other hand, we have pushed the nonlinearity from quadratic terms to cubic ones (and the procedure could be repeated), so the normalized equation has the advantage that its solution stays near the solution of the linearized equation for a longer time: although looking much worse, the

[19]The reader can check this, knowing that to eliminate xy with a Poincaré transformation we must change coordinates by $x = \xi + (\xi\eta)/3$, $y = \eta + (\eta^2 - \xi^2)/3$.

normalized equation is "simpler" in that its behavior is, for short enough times, nearer to that of the linear part.

Just to avoid any misunderstanding, we also mention explicitly that the normal form dynamic can be quite complicated and possibly chaotic [118,187,322, 323]; see also [320,335].

It should be mentioned that normal forms are also strongly related (see e.g. the discussion in [12,69,319] about the *averaging* theory and method; this is discussed at an elementary level, e.g., in [164,319] and at a more advanced one, e.g., in [3,8,12,13,17,56,69,86,225,248,252,253,285,321].

5. Normal Forms and Higher-Level Perturbation Theory

Normal forms are the basic tools of perturbation theory, together with averaging; in a way, they are the "assembler" language with which more sophisticated and advanced applications and languages can be built up. In this volume we will not deal with such "higher-level applications" of perturbation theory, but we will briefly mention them in this section, again to provide motivation and perspective for the reader.

First of all we note that normal forms theory also has an important role in *bifurcation theory* (and in singularity theory as well). The idea is that putting the bifurcation equation (which is finite-dimensional) in normal form allows us to obtain, on the one hand, a classification of the elementary bifurcations and, on the other, a better quantitative computation of the bifurcating solutions. We will not discuss these themes at all, and just refer, for example, to [3,12,14,190].

It is interesting to note that, as already mentioned before, it was essentially through *equivariant local bifurcation theory* that symmetry considerations entered into perturbation theory for dynamical systems.[20] Attention was called to this by Ruelle in connection with the degeneration encountered in the bifurcations considered by Landau theory of turbulence. Ruelle remarked that a degeneration entirely due to symmetry can be dealt with by using symmetry theory [278]. This point of view was then developed into a coherent theory first by Sattinger [288,289] (see also [236]) and then by Golubitsky and Stewart [169], and it is nowadays one of the most popular themes of investigation – theoretical and experimental, the latter in connection with *pattern formation* [30,58–60,80,99,100,105,106,169,228] – in nonlinear dynamics.

We also mention that the normalization procedure can be seen (up to the order where resonances enter into play) as a way to extend to nonlinear equations the properties of linear ones, and in particular of linear *normal modes* of oscillation; the concept of a *nonlinear normal mode* was first introduced by Weinstein [338]; see also [241,242].

[20] And it was also in this way, i.e. through the work of mathematicians, that the attention of physicists, already well aware of the role of symmetries in perturbation theory for quantum mechanics, was called to its role in the case of dynamical systems. This remark is maybe needed, should the (physicist) reader think that the point has always been clear to physicists.

Physics is particularly interested, in many domains, in *Hamiltonian systems.* If we have an Hamiltonian[21] $H = H(p,q)$ this also gives us the Hamilton equations

$$\begin{aligned} \dot{q} &= \partial H/\partial p \\ \dot{p} &= -\ \partial H/\partial q \ , \end{aligned} \tag{16}$$

so that we can apply on these the Poincaré normal forms theory, and normalization procedure. Note that we would like to consider transformations which preserve the Hamiltonian nature of the equations in particular.

It was observed by Birkhoff that one can indeed perform the Poincaré normalization and preserve the Hamiltonian structure.[22] Moreover, he also observed that in this case it is much more convenient to deal directly with the Hamiltonian, i.e. one scalar function, rather than with the vector field it generates, i.e. the $2n$ Hamilton equations.

Birkhoff also proved the exact equivalent of the Poincaré theorem for Hamiltonians, i.e. that if the quadratic part $H_0(p,q)$ of $H(p,q)$ at an equilibrium point (which for ease of notation we assume to be the origin) satisfies a *non-resonance condition*, then we can formally normalize H step by step and push the nonlinearity up to any desired order: i.e. in new coordinates (π,χ) we have $H = H_0(\pi,\chi) + H_N(\pi,\chi)$ where H_N contains terms of order N and higher, and N can be made as high as required.

The resonant case was then studied by Gustavson [174], who obtained the equivalent of the Dulac theorem in the Hamiltonian frame. The topic of Hamiltonian (or Birkhoff-Gustavson) normal forms is dealt with at length in the main body of the volume, and thus we will not discuss its basics in more detail here, briefly discussing instead its extensions and applications.

In particular, one is often interested in the case

$$H_0(p,q) \ = \ \sum_{k=1}^{n} \frac{m_k}{2} \left(p_k^2 + \omega_k^2 q_k^2\right) \ , \tag{17}$$

and we have n conserved quantities for the linearized equations around the equilibrium point, i.e. the partial Hamiltonians (energies) $I_k = (p_k^2 + \omega_k^2 q_k^2)$, the motion taking place on n-dimensional tori $\mathbf{T}^n \to \mathbf{R}^{2n}$, which are the common level sets of $\{I_1, ..., I_n\}$. It is then natural – and important in applications – to ask if these, or maybe a modification of them, are conserved within a small enough error for a long enough time. One could also ask – being optimistic – if the full nonlinear system still admits, at least near enough to the origin, n

[21] For ease of discussion, we consider the phase space to be $\mathbf{R}^{2n}$; see, e.g., [2,13,15,17,116, 230,232] for the general case on symplectic manifolds and more detail on this simplest case as well. Hamiltonian dynamical systems are also the subject of the reprint volume [227].

[22] It is now understood that this is not a special feature of Hamiltonian systems, and extends to many other kinds of systems with an additional structure, e.g. to *reversible* systems [64,145,186,213,294,295], or systems with a k-symmetry [211,212].

independent and exactly conserved quantities, or at least if some of the invariant manifolds (n-dimensional tori) for H_0 survive, maybe with a deformation, as invariant manifolds for the full Hamiltonian H.

The reader has probably recognized in the above vague formulation two of the basic problems considered – and to a large extent solved – by Hamiltonian dynamics in this century; the second is the subject of the celebrated KAM theory, stating perpetual conservation of certain invariants, corresponding to tori whose frequencies vector $\omega_k = \omega_k(I)$ satisfies certain arithmetic conditions. The reader is referred, for example, to [13,17,47,227], for the theory, and to [223,227], for some of its classical applications. Recent applications include [76,111,112,233,244]. We point out that recently KAM theory has also been approached via diagrammatic techniques originating in quantum field theory; see [42,79,152–154,156–158].

As for the first problem mentioned above, the most advanced general result in this direction is provided by Nekhoroshev theory [255], identifying conditions under which the constants of motion for the linearized problem H_0 are conserved within an error ε for times of the order $t \approx \exp[1/\varepsilon]$; this is similar to what we have seen in the previous section, but now we can go "beyond all orders", i.e. to exponential time scales. We will not discuss this theory as well, and just refer to [254,255] and to [48,163,271] for its foundations and, for example, to [41–45,49,50,77,124,160,175,176,256,257] for recent applications.

It should be mentioned that these spectacular results of perturbation theory were originally obtained using an approach not directly related to normal forms and "classical perturbation theory", but recent work by Giorgilli and collaborators showed that they can also be obtained from classical theory; the reader is referred to their papers [161,162,245] for detail.

We are not aware of any work dealing with Nekhoroshev or KAM theory in the presence of symmetry via symmetric normal forms theory. Actually, as far as we know, the presence of symmetry introduces in these cases a degeneration which must be artificially removed by splitting symmetric terms between the unperturbed and the perturbation terms. It would be interesting to obtain a direct symmetric formulation and result, confirming the statement of Stewart quoted in footnote 10 above [308].

6. Direct Applications of Normal Forms Theory in Physics

We hope the title of this section will not be misleading. Indeed, by now nonlinear dynamics has a great wealth of applications in physics, and in extremely different branches of it, and all of perturbative nonlinear dynamics can be seen as an application of normal forms.

Thus, in this section we just want to briefly mention some of the "direct" applications of normal forms; we are sure we are overlooking – among other things, due to our ignorance – other equally direct applications. This section should not be seen as a list of subject areas in which normal forms can be applied, but rather as a list of those in which it is already applied routinely enough for us to know about this application; attention should rather be on "missing fields", i.e. on fields and topics not mentioned here, and in which the application of normal forms could actually be beneficial.

One area in which normal forms are widely applied is, of course, classical mechanics. A list of references would be too long, and we will just refer the reader to the bibliographies of standard books, such as [12–14,173,223].

A closely related area is that of nonlinear oscillations (not necessarily of mechanical nature); in this respect we would like to quote, together with the seminal book by Bogoljubov and Mitropolski [56] and again [12,13,14,173, 190,223], the engineering-oriented book by Nayfeh [251], the excellent undergraduate textbooks by Glendinning [164] and Verhulst [319], and for applications to electrical oscillations in circuits the two-part review by Chua and Kokubu [83,84]. Once again the reader is actually referred to the extensive bibliographies which can be found in these references.

We have already mentioned the close relations existing among normal forms and bifurcations; see Sect. 5 and references therein, in particular [80,169,190].

It should be mentioned that recent results by L. Arnold and Imkeller [10] extended normal forms theory to the realm of *stochastic* differential equations (a topic we will not deal with at all, unfortunately, in this volume), thus opening their application to all other classes of physical problems.

Several attempts are in progress concerning the extension/application of normal forms theory (and the perturbative techniques and results based on it, such as Nekhoroshev and KAM theories) to infinite-dimensional systems, i.e. infinite chains or PDEs; we quote here in this respect only a few papers, i.e. [20–23,26,27,33,46,57,64,103,201,208,259,275,337,344], apologizing to other authors; see also [321] for a review of available results concerning averaging in PDEs.

Other attempts, dealing with different but related concepts of normal forms for PDEs include [58,59,99,100,155], with an approach based on diffusive-type equations, and [110,205,343], with an approach based on the theory of nonlinear integrable evolution equations.

A field in which normal forms are routinely used to perform high-precision computations is that of celestial – including satellite – dynamics; this problem is at the origin of a vaste literature, mainly in *celestial mechanics.*

A closely related field, where one also needs high-precision computations performed by applying normal forms theory is that of accelerator machines[23] as used in high energy physics; again we keep to a few recent references [35–37,113,115,184,311,313].

Normal forms have recently also been applied in chemistry, in particular in relation to the computation of molecular spectra [1,281–283,303]; note that in this case the symmetry plays a very relevant role.

A "quantum" version of (Hamiltonian) normal forms has also been considered, based on the fact that the Hamiltonian in Birkhoff normal form presents

[23]This is strongly related to the mathematical subject of symplectic mappings.

favourable properties for quantization; keeping again to recent works, see, e.g., [6,177,178,234,258,283,284,318].

As a rather surprising field to which normal forms have also been applied, we mention the management of complex electrical power systems and distribution networks; see [196,197,224,327].

Other applications will be mentioned where appropriate in the following chapters. Last but not least, we would like to mention the two proceedings volumes of the workshops on "symmetry and perturbation theory" [25,109], which contain a wealth of useful information and results on the subjects discussed here.

I. Symmetry and Differential Equations

In this chapter, we set a geometric framework for the study of differential equations and their symmetries. We reconduct differential equations to manifolds in an appropriate space, the *jet space.*[1] The main difference from the familiar case of algebraic equations in R^n are the relations existing between a function and its derivatives. In geometrical terms, these are taken care of by the natural *contact structure* with which the jet space is equipped.

Once a differential equation is understood as a geometric object, it is easy to define its symmetries as operations – in particular, in the Lie approach, diffeomorphisms – leaving this object invariant. The compatibility of diffeomorphisms of jet space with the contact structure leads to the requirement that they are the *prolongations* of diffeomorphisms of the base space, i.e. of dependent and independent variables. Clearly these induce an action on the space of functions and on that of derivatives, by means of which they are prolonged (lifted) to the full jet space.

We will assume in the following that the reader is somewhat familiar with the language and the basic concepts of differential geometry and group theory[2] (we give however a collection of relevant facts concerning Lie algebras at the end of Chap. IV); the less standard concepts and ideas will be introduced in this chapter or when we need them later on.

The subject of symmetry theory of differential equations was started by Sophus Lie himself, and actually was at the basis of his motivation for studying continuous groups of transformations. However, in part because of the enormous complexity of the computations needed to determine the complete list of symmetries of a given differential equation; this approach was neglected over several decades. Interest grew again thanks to G. Birkhoff, especially for fluid dynamical applications [52], and later through the Ovsjiannikov school in the USSR [265,266] and Cole and Bluman [53,54] in the USA.

In more recent years, the symmetry approach has attracted formidable interest and has proved very effective both for general theoretical analysis and

[1] Jet spaces were introduced by Ehresmann in the 1950s [120]; see e.g. [11,12,291]. The spaces of variables and their derivatives – very similar indeed to jet spaces – were already used by Lie, who introduced prolongations of vector fields.

[2] These are, fortunately, subjects on which a wide choice of excellent textbooks is available; we quote [117,192,304] for differential geometry (with a compact and very good introduction given, e.g., in [13]) and [179,203,238,239] for group theory (here again a compact and excellent introduction is provided, e.g., in [214,215]).

for obtaining special solutions; this revival – and the true progress made – was boosted by at least two circumstances: On the computational side, the appearance of symbolic manipulation programs (such as LISP, MACSYMA, REDUCE, and more recently MAPLE and MATHEMATICA) allowed to effectively implement the huge but completely algorithmic computations required by Lie's approach; on the theoretical side, the book – now already a classic – by P. Olver [262] provided a firm foundation, a clear and precise introduction, pleasant reading, and a guide for further developments.

Mentioning the main contributors to the development of the theory and its applications becomes, at this point, quite difficult, and the following quotations necessarily reflect, at least in part, the preference of the present authors. It seems, however, necessary to mention again the school of G. Bluman [55] for what concerns the "applied mathematics" side, H. Stephani [305] and his collaborators for the application to (nonlinear) equations of general relativity, and P. Winternitz and his collaborators [341] for the development of the algebraic approach and for boosting the study of conditional symmetries and of symmetries for difference and differential-difference equations.[3]

At the same time, one should mention the activity in the (former) Soviet Union of several groups and individuals, among which are N.H. Ibragimov, W.I. Fushchich, A.V. Mikhailov, A.B. Shabat, V.V. Sokolov, A.M. Vinogradov, and E.M. Vorob'ev, see for example [141,142,188,237,301,302,326,328,329].

Among recent major contributions, we would like to point out again those by P.J. Olver [263] and P. Winternitz [342].

Although we mention only works that focus specifically on the symmetry approach to differential equations, there are endless related works in nearby fields; in particular, those focused on differential equations or Lie algebras, or on integrable systems, finite and infinite-dimensional, and on classical (Lagrangian and Hamiltonian) mechanics and field theory.

From this, partial, list it is clear that even a brief summary of the different developments of the theory[4] – not to mention applications and/or related work in other fields ! – is well beyond the scope of this introduction. More modestly, we only aim to provide an introduction to the basic features of the theory that will be needed in the following discussion, and to show the beautiful geometrical ideas which underlie the algebraic and analytical setting. This first chapter relies heavily on the introductory chapter of [146], and conforms to the approach and notations of [262].

---

[3] These are topics that we will, unfortunately, not discuss in this volume. For the general approach to conditional symmetries see [218,220,222,341], and [272,273] for the relation to similar approaches. Symmetries of difference and differential-difference equations have been discussed by a number of authors, also in connection with numerical schemes; see e.g. [114,151,218,221] and references therein.

[4] A large bibliography of works published before 1993 is given in [146].

For an alternative concise introduction, the reader is referred e.g. to [341] or [188], while for a more complete treatment we simply refer to the book by Olver [262].

For the reader whose expertise and motivation lie in the theory of differential equations, it is advisable to also go through the sections on symmetry in [11].

1. Geometrical Setting

It is customary to identify an algebraic equation[5] in $\mathbf{R}^n$ and its solution set, i.e. the equation $f(x) = 0$ and the set $S_f = f^{-1}(0) = \{x : \ f(x) = 0\}$; for suitably smooth f, S_f is an embedded smooth submanifold of $\mathbf{R}^n$.

Pretty much the same identification between an algebraic relation and a geometric object can be done in the case of a differential equation Δ. This can be seen as a relation between dependent variables u, independent variables x, and derivatives – up to some order n – of the u with respect to the x. We write Δ as $F[x, u; u^{(n)}] = 0$, and assume here that F is a suitably smooth function of its arguments and, for the sake of simplicity, that $u : X \to U$, with $X \subseteq \mathbf{R}^q$, $U \subseteq \mathbf{R}^p$. For the time being, we will not discuss boundary conditions, i.e. the specification of the function space in which the differential equation is defined.

The geometrical object to which Δ is identified is again called a solution set $S_\Delta = F^{-1}(0)$, but this is now a subset not of some $\mathbf{R}^k$ space, but of the appropriate *jet space* (of the same order n as the differential equation). This jet space is a generalization of the tangent bundle $\mathrm{T}M$ of a manifold M, and is similar to the familiar phase space; our first task will be to understand the jet space.

Let us first recall that the tangent bundle [2,13] is obtained by associating with any point $x \in M$ the tangent space to M at x, $\mathrm{T}_x M$. If M is d-dimensional, it is locally isomorphic to $\mathbf{R}^d$, and $\mathrm{T}_x M$ is itself $\mathbf{R}^d$. It should be stressed that spaces tangent to different points of M, although isomorphic, should not be confused, and that a priori we do not have any preferred way of assigning a correspondence among two such spaces.

In the simplest case of $M = \mathbf{R}^d$, the tangent bundle is given (like the phase space) by $\mathrm{T}M = \mathbf{R}^d \times \mathbf{R}^d$. In physical terms, this means that a point in $\mathrm{T}M$ is specified by a point on M and a velocity (tangent) vector at that point. We do not have, a priori, any way of comparing velocity (tangent) vectors at different points of M.

Recall that a tangent vector to M at x is also defined as a class of equivalence of curves $s(t) \in M$ through $x = t_0$, two such curves $s(t)$ and $r(t)$ being equivalent if for $t \to t_0$ we have $|s(t) - r(t)| = o(t)$; the fiber bundle could now be defined as the semidirect product of M and these equivalence classes of curves through

---

[5] By this we will mean a relation between variables $x_1, x_2, ... x_n$ of the form $f(x) = 0$, i.e. not involving derivatives of the variables, even if f is not an algebraic function.

points of M. As tangent spaces to different points of M are isomorphic, so are the equivalence classes of curves, and thus TM is topologically a direct product.

In the same way, we can define a *jet of order* p at $x \in M$ as an equivalence class of curves $s(t) \in M$ through $x = t_0$, two such curves $s(t)$ and $r(t)$ being now equivalent if for $t \to t_0$ we have $|s(t) - r(t)| = o(t^p)$. The jet space – or *jet bundle* – is then defined in the same way as the tangent bundle, which is indeed a jet space of order one for the case of only one independent variable (i.e. the time).

We can now also define locally[6] the jet space in a less geometric way, using explicitly the coordinates. For this, we use the natural coordinates $(x_1, ..., x_q) \in X$ and $(u_1, ..., u_p) \in U$ given by our assumption that $X \subseteq \mathbf{R}^q$, $U \subseteq \mathbf{R}^p$.

We can consider the space $U_{[1]} \subseteq \mathbf{R}^{q \cdot p}$ of first derivatives of u's with respect to x's, with basis[7] $(\partial_1 u_1, \partial_2 u_1, \dots, \partial_q u_p)$. We can also, analogously, consider the spaces $U_{[k]}$ of k-th order derivatives of u's with respect to the x's; for this we have $U_{[k]} \subseteq \mathbf{R}^{N_k}$, with $N_k = (p+k-1)!/[(p-1)!k!]$ (the ordering of derivatives is not important).

We will find useful to use the *multiindex* notation: $J = (j_1, \dots, j_q)$ is a multiindex, and $u_J^{(i)} = \partial_1^{j_1} \dots \partial_q^{j_q} u^{(i)}$. The *order* of J is $|J| = \sum_{i=1}^q j_i$, so that $u_J \in U_{[|J|]}$.

Now, let $M = X \times U$; the m-th *jet space of M*, or the jet space of order m, is locally

$$M^{(m)} = X \times U \times U_{[1]} \times \cdots \times U_{[m]} \ . \tag{1}$$

This is also called the m-th prolongation of M. Analogously, the m-th prolongation of U is locally

$$U^{(m)} = U \times U_{[1]} \times \cdots \times U_{[m]} \tag{2}$$

and, of course, we have locally

$$M^{(m)} = X \times U^{(m)} \tag{3}$$

We also have locally $U^{(m)} \approx \mathbf{R}^{P_m}$, $P_m = \left(\prod_{k=1}^m N_k\right) \cdot p$. For ease of notation, in the following we will also use

$$U_{[0]} \equiv U \quad ; \quad N_0 = p$$

It should be quite clear that the m-th jet space provides a natural frame for viewing a differential equation of order m as a geometrical object.

We have seen that the m-th jet space is locally isomorphic to a real space $\mathbf{R}^{P_m}$. However, there is a natural structure in it, called *contact structure* [12], that embodies the peculiar relations among the $U_{[k]}$.

This is easily understood by considering the case $q = p = 1$, so that $X \simeq \mathbf{R}$, $U \simeq \mathbf{R}$, $M \simeq \mathbf{R}^2$, and consider the first jet space $M^{(1)} = X \times U \times U_{[1]}$ with coordinates (x, u, v). If now we have a curve $\gamma \in M$ corresponding to a function

[6] These local definitions also hold globally if – as is often the case – $X = \mathbf{R}^q$, $U = \mathbf{R}^p$.

[7] Here and in the following $\partial_i \equiv \partial / \partial x_i$.

$u(x)$ with derivative $\mathrm{d}u/\mathrm{d}x = v(x)$, and we consider the curve $\Gamma \in M^{(1)}$ given by $(x, u(x), v(x))$, the curve Γ is not an arbitrary curve in $\mathbf{R}^3$, but must satisfy, at any point, the condition $\mathrm{d}u(x)/\mathrm{d}x = v(x)$.

In other words, any curve in $M^{(1)} \simeq \mathbf{R}^3$ must be everywhere tangent to a field of planes in $\mathbf{R}^3$, identified by the above condition. Equivalently, we can define a one-form ω, called the *contact form* and given in this case by

$$\omega = \mathrm{d}u - v\mathrm{d}x , \tag{4}$$

which must be identically zero on Γ.

More generally, if the basis in $M^{(m)}$ is given by $((x_1, \ldots, x_q), (u_1, \ldots, u_p); v_J^1, \ldots, v_J^p)$, where $v_J^i \in U_{[|J|]}$, we have a set of equations

$$\frac{\mathrm{d}u_i}{\mathrm{d}x_j} = v_j^i \quad , \quad \frac{\mathrm{d}^2 u_i}{\mathrm{d}x_j \mathrm{d}x_k} = v_{jk}^i \quad , \quad \ldots \quad , \quad \frac{\mathrm{d}^m u_i}{\mathrm{d}x_1^{j_1} \cdots \mathrm{d}x_q^{J_q}} = v_J^i \ , \tag{5}$$

defining a field of (hyper) planes; this is the contact structure of the jet space. A vector field (or a manifold) in $M^{(m)}$ is compatible with the contact structure if it is everywhere tangent to this field of (hyper) planes.[8]

Notice that it is always possible to project a manifold $\gamma(x) \in M^{(m)}$ compatible with the contact structure to a manifold[9] $s(x) = \pi[\gamma(x)] \in M$; conversely, given a smooth manifold $s(x) \in M$ in the base space, this can be lifted to the m-th jet space $M^{(m)}$ (for any m); by requiring the lifted manifold $\gamma(x) = \pi^{-1}[c(x)] \in M^{(m)}$ to be compatible with the natural contact structure in $M^{(m)}$, this lifted manifold is unique.[10]

Now, let us consider a differential equation Δ of order m, given by

$$F(x, u^{(m)}) = 0 \tag{6}$$

with F a smooth function of its arguments. As anticipated, we will identify the differential equation and its *solution manifold* $S_\mathcal{D} \subset M^{(m)}$. Notice that by doing this we have identified differential equations Δ and $\widetilde{\Delta}$ which define the same solution manifold.[11] We stress that Δ should at this stage be seen as an algebraic equation in $M^{(m)}$, with the u_J being independent variables, as their relation is taken into account by the contact structure.

If we look for a smooth solution $u = f(x)$ of (1), this will be a function

$$f : X \to U \ . \tag{7}$$

---

[8] Equivalently, this could be defined in terms of a (set of) differential form, the contact forms, which should vanish identically.

[9] Note that this projection could take smooth manifolds into non-smooth ones, as can be seen e.g. by considering solutions of the ordinary differential equation $u_x^2 = x$.

[10] This statement is equivalent to saying that once we have given the smooth function $u(x)$, we also have all its partial derivatives, see below.

[11] Such as, e.g., $u_t - u = 0$ and $(u_t - u)^2 = 0$.

This function defines a graph $\Gamma(f)$ in M

$$\Gamma(f) = \{(x,u) \in M : \ u = f(x)\} \subset M \tag{8}$$

and once again we identify an analytic object (the function f) and the corresponding geometrical one (the manifold $\Gamma(f)$).

Quite obviously, once we have $u = f(x)$, we also have all of its partial derivatives, so $\Gamma(f)$ can be lifted, as discussed above, to a manifold in $M^{(m)}$:

$$\Gamma^{(m)}(f) = \{(x, u^{(m)}) \in M^{(m)} : \ u = f(x)\} \subset M^{(m)} \ ; \tag{9}$$

this is automatically compatible with the contact structure.

Analogously, we can define the prolongation of f by associating to $u = f(x)$ the various partial derivatives; if we get up to derivatives of order m, we have the m-th prolongation of f, denoted $f^{(m)}$. This is a function

$$f^{(m)} : X \to U^{(m)} \tag{10}$$

and its graph is

$$\Gamma(f^{(m)}) = \{(x, u^{(m)}) \in M^{(m)} : \ u^{(m)} = f^{(m)}(x)\} \subset M^{(m)} \ . \tag{11}$$

By construction, this is compatible with the contact structure, and moreover

$$\Gamma(f^{(m)}) = \Gamma^{(m)}(f) \ . \tag{12}$$

If we have a solution $u = f(x)$ of (1), obviously

$$\Gamma(f^{(m)}) \subset S_{\mathcal{D}} \tag{13}$$

Conversely, solutions are geometrically obtained by considering the restriction of the hyperplane field defining the contact structure to the solution manifold; this defines a field of $(d-1)$-planes (if the contact structure is defined by a field of d-planes); the solutions are then obtained by constructing *integral submanifolds* for this field, i.e. submanifolds everywhere tangent to the planes of the field. Notice that in the case of an ordinary differential equation we would have a genuine line field on $S_{\mathcal{D}}$, and correspondingly solutions would be *integral curves* for this line field.

This approach is closely related to the classical Frobenius theorem, which we will not discuss here; for a modern discussion see e.g. [11–13,262,263].

2. Invariance of Equations and of Solutions

We want now to consider the action of groups – and in particular groups of diffeomorphisms – on differential equations and their solutions. This task will be made easier by our previous construction, as we, actually, have to consider action on submanifolds of $M^{(m)}$.

We will denote the group of diffeomorphisms of M by $\mathrm{Diff}(M)$; its algebra is denoted by *diff* $(M) = \mathcal{M}$ and is the (infinite-dimensional) Lie algebra of vector fields on M, equipped as usual with the commutator bracket $[.,.] : \mathcal{M} \times \mathcal{M} \to \mathcal{M}$ given by

$$[X_1, X_2] = X_1 X_2 - X_2 X_1 \ . \tag{14}$$

If the vector fields X_1, X_2 are written in local coordinates as

$$X_1 = f^i(x)\partial_i \quad , \quad X_2 = g^i(x)\partial_i \ , \tag{15}$$

then the vector field $Y = [X_1, X_2]$ is given in local coordinates by $Y = h^i(x)\partial_i$, where

$$h^i = \{f, g\}^i \ \equiv \ \left(f^j \cdot \partial_j\right) g^i - \left(g^j \cdot \partial_j\right) f^i \ . \tag{16}$$

In our case, the manifold M will be the base space for a differential equation, i.e. $M = X \times U$. In order to single out the actions on X and U, we will often use the notation

$$X = \xi^i(x, u)\partial_{x^i} + \varphi^j(x, u)\partial_{u^j} \ , \tag{17}$$

which we also write for short as

$$X = \xi\partial_x + \varphi\partial_u \ . \tag{18}$$

However, as Δ is identified with a manifold in $M^{(m)}$, we should also consider transformations of $M^{(m)}$. These will give Diff$(M^{(m)})$; the corresponding algebra *diff*$(M^{(m)})$ will be composed of vectors (with obvious notation)

$$Z = \xi^i(x, u^{(m)})\partial_{x^i} + \varphi^i(x, u^{(m)})\partial_{u^i} + \Phi^{(j)}_\alpha(x, u^{(m)})\partial_{u^\alpha_J} \tag{19}$$

This is too general for our purposes, as it does not take into account the restrictions given by the contact structure. We should therefore consider the group of diffeomorphisms of $M^{(m)}$ compatible with (i.e., which preserve) the contact structure. These will be denoted by Diff$^{(m)}(M)$, with algebra *diff*$^{(m)}(M)$, and will correspond to vector fields Z which are the prolongation of vector fields X on M. We also write $X^{(m)}$ for the m-th prolongation of the vector field X.

This puts a precise restriction on (1): in particular, once the ξ's and φ's are given, all the corresponding Φ's are automatically determined.[12]

It is important to note that if we have a Lie algebra $\mathcal{G}$ of vector fields spanned by $\{X_1, ..., X_n\}$ and we consider the prolongations $\{Z_1, ... Z_n\}$ of these, the Z_i also span a Lie algebra $\mathcal{G}^{(m)}$, which is isomorphic to $\mathcal{G}$, i.e. has the same commutation relations:

$$[X_i, X_j] \ = \ c^k_{ij} \ X_k \iff [Z_i, Z_j] \ = \ c^k_{ij} \ Z_k \ . \tag{20}$$

If we consider the projection of Z on the directions tangent to M, we get a vector field on M,

$$\pi\,(Z) = \xi^i(x, u^{(m)})\partial_{x^i} + \varphi^i(x, u^{(m)})\partial_{u^i} \ . \tag{21}$$

[12]The explicit expression of the Φ in terms of ξ, φ is given by the prolongation formula, see below.

We say that $\pi(Z)$ is a *Lie-point*, or *geometric*, vector field, if ξ and φ depend only on coordinates in M, i.e.

$$\pi(Z) = \xi^i(x,u)\partial_{x^i} + \varphi^i(x,u)\partial_{u^i} \; ; \tag{22}$$

and that it is a *generalized* vector field if ξ and φ can depend on u derivatives.

Obviously, only Lie-point ones correspond to genuine geometrical transformations on M, and we will mainly be interested in this kind of transformations. It should, however, be stressed that generalized symmetries (corresponding to generalized vector fields) play a relevant role in the theory of integrable systems. They will be discussed in later chapters, and also we refer readers to [262].

The expression for the $\Phi_\alpha^{(j)}$ in terms of ξ and φ is given by the *prolongation formula*; which will be derived in the Appendix at the end of this chapter and is given by

$$\Phi_\alpha^{(J)} = D_J(\varphi^\alpha - \sum \xi^i u_i^\alpha) + \sum \xi^i u_{i,J}^\alpha \; , \tag{23}$$

where we have introduced the total (or, variational) derivative operators

$$D_i = \partial_i + \frac{\partial u^\alpha}{\partial x^i}\partial_{u^\alpha} + \frac{\partial^2 u^\alpha}{\partial x^i \partial x^h}\partial_{u_h^\alpha} + \cdots \; . \tag{24}$$

Having understood the action of vector fields in the jet space, let us now go on to consider invariance of functions and later of equations.

Consider the manifolds $X \subseteq \mathbf{R}^q$, $U \subseteq \mathbf{R}^p$, and a function $f : X \to U$ which we identify with its graph $\Gamma(f)$, where

$$\Gamma(f) = \{(x,u) \in M = X \times U : \; u = f(x)\} \subset M \; . \tag{25}$$

Now, given the group $G = \mathrm{Diff}(M)$, we can consider its subgroup G_f which leaves $\Gamma(f)$ (globally) invariant, i.e.

$$G_f = \{g \in \mathrm{Diff}(M) : \; g : \Gamma(f) \to \Gamma(f)\} \subset \mathrm{Diff}(M) \; . \tag{26}$$

The corresponding Lie algebra will be

$$\mathcal{G}_f = \{X \in \mathit{diff}(M) : \; X : \Gamma(f) \to \mathrm{T}\Gamma(f)\} \subset \mathit{diff}(M) \; ; \tag{27}$$

in fact, $\Gamma(f)$ is invariant under the flow generated by X if and only if X is tangent to $\Gamma(f)$ at all points of $\Gamma(f)$.

The G_f (respectively, $\mathcal{G}_f$) is called the symmetry group (respectively, algebra) of the differential equation Δ identified by f, or – for short – of f. If we want to stress the fact that it is a subgroup of $\mathrm{Diff}(M)$ [a subalgebra of $\mathit{diff}(M)$], we will call it the isotropy subgroup (subalgebra) of f.

Consider now a given vector field

$$X = \xi^i \partial_i + \varphi^j \partial_{u^j} \; , \tag{28}$$

and the manifold $\Gamma(f)$. Let the one parameter subgroup generated by X be

$$T_\vartheta : (x,u) \to (\widetilde{x},\widetilde{u}) = \; (g_\vartheta(x,u) \quad , \quad h_\vartheta(x,u)) \; . \tag{29}$$

This has to be an invertible transformation by the very definition of a group; its inverse will be

$$T_{-\vartheta} : (\widetilde{x}, \widetilde{u}) \to (x, u) = \ (g_{-\vartheta}(\widetilde{x}, \widetilde{u}), h_{-\vartheta}(\widetilde{x}, \widetilde{u})) \ \equiv \ (\gamma_\vartheta(\widetilde{x}, \widetilde{u}), \eta_\vartheta(\widetilde{x}, \widetilde{u})) \ ; \qquad (30)$$

obviously, here we have

$$\left[\frac{\mathrm{d}g_\vartheta^i(x,u)}{\mathrm{d}\vartheta}\right]_{\vartheta=0} = \xi^i(x,u) \quad , \quad \left[\frac{\mathrm{d}h_\vartheta^j(x,u)}{\mathrm{d}\vartheta}\right]_{\vartheta=0} = \varphi^j(x,u) \ . \qquad (31)$$

Therefore, a point in $\Gamma(f)$ is transformed according to

$$T_\vartheta : (x, f(x)) \to \ (g_\vartheta(x, f(x)), h_\vartheta(x, f(x))) \ , \qquad (32)$$

and this is by definition the point $\left(\widetilde{x}, \widetilde{f}_\vartheta(\widetilde{x})\right)$. In this way we define an induced action $\widetilde{T}$ of Diff(M) on the space of smooth functions $f : X \to U$. This is given by

$$\widetilde{T}_\vartheta : f \to \widetilde{f}_\vartheta \ , \qquad (33)$$

corresponding to $T_\vartheta : \Gamma(f) \to \Gamma(\widetilde{f}_\vartheta)$. In this way we have defined

$$\widetilde{f}_\vartheta(\widetilde{x}) = h_\vartheta(x, f(x)) \qquad (34)$$

and, in terms of infinitesimal transformation,

$$\widetilde{f}_\varepsilon(\widetilde{x}) = f(x) + \varepsilon\varphi(x, f(x)) \ . \qquad (35)$$

Using now $\widetilde{x} = x + \varepsilon\xi + O(\varepsilon^2)$, we get

$$\widetilde{f}_\varepsilon^i(\widetilde{x}) = f^i(\widetilde{x}) + \varepsilon \left[\varphi^i(\widetilde{x}, f(\widetilde{x}) - \xi^j(\widetilde{x}, f(\widetilde{x})) \cdot \frac{\partial f^i(\widetilde{x})}{\partial \widetilde{x}_j}\right] \ . \qquad (36)$$

We also write that $\widetilde{f} = f + \varepsilon\delta f$, where $\delta f = \delta u - (\partial f/\partial x)\delta x$; thus the generators which leave f invariant, i.e. for which $\delta f = 0$, are those which yield

$$\delta u = \frac{\partial f}{\partial x}\delta x \ . \qquad (37)$$

In terms of φ and ξ, this reads $\varphi^i(x, f(x)) = \left(\xi^j(x, f(x)) \cdot \partial_j\right) f^i(x)$, so that we have a characterization of $\mathcal{G}_f$ as

$$\mathcal{G}_f = \{X = \xi\partial_x + \varphi\partial_u : \ \varphi^i = \left(\xi^j \cdot \partial_j\right) f^i\} \ . \qquad (38)$$

We have thus described the symmetries of a function f by studying the invariance of the corresponding manifold Γ_f; we can in the same way consider the invariance of an equation, as this is identified with a manifold. Let us first consider the action of Diff(M) on an algebraic equation Δ, which we write in the form $F(x,u) = 0$. As it was discussed before, Δ identifies a solution manifold S_F in M; notice that this corresponds to a level surface of F.

If we are given a vector field $X \in diff(M)$, this will generate a one-parameter subgroup T_ϑ of Diff(M); in general $T_\vartheta : (x,u) \to (g_\vartheta(x,u), h_\vartheta(x,u))$, so that

$$T_\vartheta : S_F \to T_\vartheta(S_F) = \{(x,u) \in M : \; F(g_{-\vartheta}(x,u), h_{-\vartheta}(x,u)) = 0\} \tag{39}$$

The condition needed in order to have S_F invariant under T_ϑ, i.e. $T_\vartheta : S_F \to S_F$, is, in terms of the vector field, $X : S_F \to \mathrm{T}S_F$; we define the symmetry group of F as

$$G_F = \{g \in \mathrm{Diff}(M) : \; g : S_F \to S_F\} \; , \tag{40}$$

and its algebra is

$$\mathcal{G}_F = \{X \in diff(M) : \; X : S_F \to \mathrm{T}S_F\} \; . \tag{41}$$

This looks quite similar to what was done for functions, but one should pay attention to the differences: here an algebraic equation is associated with a function $F : X \times U \to \mathbf{R}$ and identified not with the graph of this function (which is a subset of $M \times \mathbf{R}$), but with the zero level set of this function (which is a subset of M).

If we have a solution $u = f(x)$ to Δ, then

$$\mathcal{G}(f) \subset S_F \; ; \tag{42}$$

if we consider the solution set of F, i.e.

$$\mathcal{S}_F = \{f : \; F(x, f(x)) = 0\} \; , \tag{43}$$

then the union of graphs of functions in $\mathcal{S}_F$ is in S_F.

Notice that the condition $X : S_F \to \mathrm{T}S_F$ means that $XF(x,u) = 0$ on S_F. Therefore, we can equivalently define

$$\mathcal{G}_F = \{X \in diff(M) : \; [XF(x,u)]_{S_F} = 0\} \; . \tag{44}$$

This discussion, conducted for an algebraic equation, is analogous for differential equations, thanks to the jet space construction.

We write a differential equation Δ of order m as

$$F(x, u^{(m)}) = 0 \tag{45}$$

and identify it with its solution manifold

$$S_F = \{(x, u^{(m)}) \in M^{(m)} : \; F(x, u^{(m)}) = 0\} \subset M^{(m)} \; . \tag{46}$$

The symmetry group G_F of Δ will be

$$G_F = \{g \in \mathrm{Diff}(M) : \; g^{(m)} : S_F \to S_F\} \; , \tag{47}$$

with symmetry algebra

$$\mathcal{G}_F = \{X \in diff(M) : \; X^{(m)} : S_F \to \mathrm{T}S_F\} \; . \tag{48}$$

With the same considerations as in the case of algebraic equations, we can equivalently define

$$\mathcal{G}_F = \{X \in \mathit{diff}(M) : \ \left[X^{(m)}F\right]_{S_F} = 0\} \ . \tag{49}$$

It should be stressed that, as is clear from the above definition of $\mathcal{G}_F$, in order to have X as a symmetry of Δ, we do *not* need to have $X^{(m)}(F) = 0$ out of S_F. If $X^{(m)}(F) \equiv 0$ on all of $M^{(m)}$, we say that X is a *strong symmetry* of Δ. The relation between ordinary and strong symmetries of differential equations has been clarified in [74].

3. Reduction and Solution of Ordinary Differential Equations

Let us now briefly discuss how knowledge of the symmetry of an ordinary differential equation (ODE) can help in obtaining its solution, or at least in reducing its order.

We consider an equation of order m for $x(t) \in R^n$, which we write as

$$F(t, x^{(m)}) \ = \ 0 \ , \tag{50}$$

and a vector field $v \in \mathcal{G}$, where $\mathcal{G}$ is the symmetry algebra of $F = 0$. If we are able to determine a change of coordinates $x = x(w,z), t = t(w,z)$ such that in the new coordinates v reads simply

$$v \ = \ \frac{\partial}{\partial w_n} \tag{51}$$

then the function

$$G(z, w^{(m)}) \ = \ F\left(t(w,z), x(w,z)\right) \tag{52}$$

necessarily satisfies $v(G) = 0$, i.e.

$$\frac{\partial G(z, w^{(m)})}{\partial w_n} \ = \ 0 \ . \tag{53}$$

This means that G does not depend on w_n, and we can therefore consider the new coordinates

$$y_i \ = \ \begin{cases} w_i & \text{for } i = 1, ..., n-1 \\ \mathrm{d}w_n/\mathrm{d}t & \text{for } i = n. \end{cases} \tag{54}$$

In this way, we reduce the equation to an equation $\mathcal{F}[z, y] = 0$ which is of order m in $y_1, ..., y_{n-1}$, and of order $m-1$ in y_n (assuming, for ease of discussion, that F was of order m in all the x_i).

The change of coordinates $x = x(w,z), t = t(w,z)$ can be determined by the method of characteristics [102]: indeed, consider the flow on M under v; if $v = \partial/\partial w_n$, the other coordinates $(z; w_1, ..., w_n)$ are constant under v. Thus, if we denote by ξ any of these coordinates, we have $v(\xi) = 0$; in order to determine the ξ as a function of $(t;x)$ we can just write v in these coordinates, so that we have a first order homogeneous PDE. In other words, if

$$v = v_0(t;x)\partial_t + \sum_{i=1}^{n} v_i(t;x)\partial_i \ , \tag{55}$$

then each of the ξ satisfies

$$v_0(t;x)\frac{\partial \xi}{\partial t} + \sum_{i=1}^{n} v_i(t;x)\frac{\partial \xi}{\partial x_i} = 0 \ . \tag{56}$$

As is well known, the solutions to this equation are given by arbitrary functions of $\zeta_1(t;x), ..., \zeta_n(t;x)$ obtained by the characteristic equation

$$\frac{\mathrm{d}t}{v_0(t;x)} = \frac{\mathrm{d}x_1}{v_1(t;x)} = \ldots = \frac{\mathrm{d}x_n}{v_n(t;x)} \ . \tag{57}$$

Thus, once we have solved the characteristic equation, we can choose any n functions of the ζ (provided that they are linearly independent at each point of M) as the z and $w_1, ..., w_{n-1}$. Clearly a natural choice is that these are just the $\zeta_1, ..., \zeta_n$ themselves. The w_n is then simply chosen by requiring it to be linearly independent of these at all points of M, or equivalently that $(\zeta_1, ..., \zeta_n, w_n)$ are a set of functionally independent functions of $(t; x_1, ..., x_n)$.

If (50) were a single equation of order m, i.e. for $n = 1$, we would have thus reduced its order by one; similarly, if (50) were a system of first order ODEs, then $\mathcal{F}$ does not contain derivatives of y_n, and the system has been separated into a system of $n-1$ first order ODEs and one algebraic equation (of the form $y_n = f(t; y_1, ..., y_{n-1})$).

It should be stressed that we can operate in this way also without knowing the *full* symmetry algebra $\mathcal{G}$, as soon as we know *one* symmetry vector field v.

We also remark that the change of coordinates $y_n = \mathrm{d}w_n/\mathrm{d}t$ is not invertible, and its inverse $w_n = \int y_n \mathrm{d}t$ depends on an arbitrary constant. This is not surprising, since when we reduce the order of the system by one we lose an arbitrary integration constant in the form of the general solution, and it is precisely this constant that appears when we invert $y_n = \mathrm{d}w_n/\mathrm{d}t$ in order to recover the solution to $F = 0$ from the solution to $\mathcal{F} = 0$.

Example 1. As a first, completely trivial, example, consider the equation

$$F\left(t, x^{(1)}\right) \equiv \frac{\mathrm{d}x}{\mathrm{d}t} + u = 0 \ , \tag{58}$$

which obviously has $v = \partial/\partial t$ as a symmetry; according to our procedure, we should express v as $v = \partial/\partial w$, which can be done via the change of coordinates $x = z$, $t = w$ (thus we just exchange the role of dependent and independent coordinates). In these coordinates, $w_z = (1/x_t)$, and the equation $F = 0$ reads, in the (z, w) coordinates,

$$G\left(z, w^{(1)}\right) \equiv \frac{1}{w_z} + z = 0 \ , \tag{59}$$

which indeed satisfies $v(G) = 0$, i.e. is independent of w. We can then go to the coordinates (z, y), where $y = w_z$, and the equation is now

$$\mathcal{F}(z,y) \equiv \frac{1}{y} + z = 0 , \tag{60}$$

i.e. we are reduced to an algebraic equation. The solution to this is $y = -(1/z)$, which yields $w = -\ln(z)+\ln(c)$ (where c is a constant), and thus $\ln(c)-t = \ln(x)$, that is $x(t) = ce^{-t}$.

Once we have performed the full (symmetry reduction) procedure described above, we are left with a reduced system of ODEs. This system will have a symmetry algebra $\widetilde{\mathcal{G}}$, and in particular will inherit some of the symmetry $\mathcal{G}$ of the original system (50). We can then apply again the same procedure using a different vector field, and so on, until we have some (known) symmetry vector field to employ.

`Example 2.` We consider now a less trivial example, having a two-dimensional symmetry algebra:

$$F\left(t, x^{(2)}\right) \equiv \frac{\mathrm{d}^2 x}{\mathrm{d}t^2} + \left[\left(\frac{1}{t}\right) - \left(\frac{x}{t^3}\right)\right]\frac{\mathrm{d}x}{\mathrm{d}t} = 0 . \tag{61}$$

This is invariant under the vector field $v = t\partial_t + 2x\partial_x$, whose second prolongation is given by $v^{(2)} = t\partial_t + 2x\partial_x + x_t\partial_{x_t}$. Notice that v generates a scaling transformation $t \longrightarrow \lambda t$, $x \longrightarrow \lambda^2 x$, so that it is obvious that x/t^2 is an invariant function for v (and x_t/t, x_{tt} are invariant functions for $v^{(2)}$); alternatively, one could obtain the same result by considering the characteristic equation. We can thus choose as coordinates $z = x/t^2$ and $w = \ln(t)$; in terms of these, $f = 0$ reads

$$G\left(z, w^{(2)}\right) \equiv w_{zz} + (4-z)\left(\frac{w_z}{1+zw_z}\right)^2 - \left(z^2 - z\right)\left(\frac{w_z}{1+zw_z}\right)^3 = 0 . \tag{62}$$

By setting $y = w_z$, we are left with a first-order equation, i.e.

$$\mathcal{F}\left(z, y^{(1)}\right) \equiv \frac{\mathrm{d}y}{\mathrm{d}z} + (4-z)\left(\frac{y}{1+zy}\right)^2 - \left(z^2 - z\right)\left(\frac{y}{1+zy}\right)^3 = 0 , \tag{63}$$

i.e. we have indeed used one symmetry to reduce the order of the differential equation by one.

Two points should be made clear concerning the procedure briefly described above (see [262,146] for a more detailed discussion). The first is that the symmetry algebra $\widetilde{\mathcal{G}}$ of the reduced equation could contain, beyond the vector fields "inherited" from the symmetry algebra $\mathcal{G}$ of the original equation, some other symmetry arising in the reduction procedure [168].

The second point concerns the relation between $\mathcal{G}$ and $\widetilde{\mathcal{G}}$. If we denote by $\mathcal{G}_1$ the (one-dimensional) algebra with generator v, we have that $\widetilde{\mathcal{G}} \subseteq \mathcal{G}/\mathcal{G}_1$. However, if $\mathcal{G}_1$ is not an ideal in $\mathcal{G}$, the set $\mathcal{G}/\mathcal{G}_1$ is not a Lie algebra, but only a lateral class; thus, $\widetilde{\mathcal{G}}$ will be given by the largest subset of $\mathcal{G}/\mathcal{G}_1$ which is a Lie algebra. Equivalently, we have that $\widetilde{\mathcal{G}} = \mathcal{G}_n/\mathcal{G}_1$, where $\mathcal{G}_n \subseteq \mathcal{G}$ is the largest solvable subalgebra of $\mathcal{G}$ which contains $\mathcal{G}_1$ as an ideal. If G is the group generated by $\mathcal{G}$, and G_1, G_n the groups generated by $\mathcal{G}_1$ and G_n, then G_n is just the normalizer of G_1 in G.

When we consider repeated symmetry reductions, the above argument shows that the effective reduction possible in this way corresponds to using the vector fields $v_1, ..., v_h$ which span a maximal *solvable* subalgebra $\mathcal{G}_s$ of $\mathcal{G}$.

It should be stressed moreover that one should consider the symmetry reductions in the appropriate order: namely, at each stage one should use a symmetry vector field which spans a one-dimensional ideal in the residual algebra, otherwise we would lose the possibility of performing further reductions. A clear example of this situation is provided in [262] (and reported in [146]).

It should be mentioned that this same procedure can also be understood in terms of differential invariants [262,263], as we now briefly explain. Given a vector field v acting in M, a geometric invariant for v is a function $\zeta : M \to R$ such that $v(\zeta) = 0$. Similarly, a differential invariant[13] of order k is a function $\zeta : M^{(k)} \to R$ such that $v^{(k)} \left(\zeta(t; x^{(k)}) \right) = 0$.

The key observation [262] is that if ζ_1 and ζ_2 are k-th order invariants, i.e. $\zeta_i = \zeta_i \left(t; x^{(k)} \right)$, $v^{(k)} \left(\zeta_i \right) = 0$ for $i = 1, 2$, then we can construct a $(k+1)$-th order invariant simply as

$$\zeta_3 \left(t; x^{(k+1)} \right) = \frac{D_t \zeta_2}{D_t \zeta_1} . \tag{64}$$

In the case of a single ODE, of any order m, this means that once we have determined the zero-th and first-order differential invariants for a vector field v, we can determine the differential invariants of any order simply by applying the above formula.

Example 3. Let us consider again the vector field $v = t\partial_t + 2x\partial_x$, with first prolongation $v^{(1)} = v + x_t \partial_{x_t}$. As remarked above, the differential invariants of order one of this are $\zeta_1 = x/t^2$ and $\zeta_2 = x_t/t$. Then,

$$\frac{D_t \zeta_2}{D_t \zeta_1} = \frac{x_{tt}/t - x_t/t^2}{x_t/t^2 - 2x/t^3} = \frac{x_{tt} - \zeta_2}{\zeta_2 - 2\zeta_1} ; \tag{65}$$

this is an invariant if and only if x_{tt} is a differential invariant (of order two), as we know indeed that it is.

We then have to recall a classical result [262], i.e. that if $F \left(t; x^{(m)} \right) = 0$ admits a symmetry vector field v, then there exists an equation $\widetilde{F} \left(t; x^{(m)} \right) = 0$ for which v is a strong symmetry (these were defined in the previous section) and which has the same solution set as $F = 0$. We can then work with the equivalent

[13] Although our notation here refers to the ODE case, this can be immediately extended to the case of PDEs.

equation $\widetilde{F} = 0$ rather than with $F = 0$; let us now assume that v is a strong symmetry of $F = 0$, for ease of notation.

Since $F\left(t; x^{(n)}\right)$ has v as a strong symmetry, F can be written as a function of the n-th order differential invariants for v, i.e. of the invariants for $v^{(n)}$. Thus, once we know the differential invariants $\zeta_1, ...\zeta_n$ of v of order m, we can express $(t; x^{(n)})$ in terms of $(\zeta_1, ..., \zeta_n)$ and of an additional variable w_n, necessarily not invariant under $v \neq 0$. As the $(\zeta_1, ..., \zeta_n)$ are invariants and all solutions of $v^{(n)}(F)$ are written in terms of these, it must be that $\partial F/\partial w_n = 0$ and in these coordinates $v = \partial/\partial w_n$, i.e. we recover the setting considered above.

4. Reduction and Solution of Partial Differential Equations

We have seen that in the case of ODEs, a knowledge of the symmetry algebra of the equation is quite expedient in determining, via a definite algorithm, a reduction of the equation itself and – if the algebra $\mathcal{G}$ contains a solvable algebra $\mathcal{G}_s$ of sufficiently large dimension – even to give the complete solution of the equation.

In the case of PDEs, we should not expect to be able to obtain such a general result. Indeed, we know that in general we are not able, in any way, to obtain the most general solution to a nonlinear PDE.

Thus, in the case of PDEs we should be less ambitious. We can still use symmetry methods, to good effects, in two ways: 1) Determine special solutions to the given PDE; and 2) once we know a special solution, obtain different solutions by applying the symmetries of the equation.

It should be mentioned that, as we shall briefly see in the following, by giving up the search for global[14] reductions of the PDE we are enabled to use not only the symmetries of the equation but, in some respect, also the symmetries of special solutions that are not at the same time symmetries of the equation (these also go under the name of *conditional symmetries*, see below).

Let us first concentrate on the second point mentioned above, and as usual denote by $\mathcal{G}$ the symmetry algebra of the equation we want to study, say

$$F\left(x, u^{(m)}\right) = 0 , \qquad (66)$$

with $x \in \mathbf{R}^q$, $u \in \mathbf{R}^p$. We know that if $v \in \mathcal{G}$, then $v : S_F \to \mathrm{T}S_F$. This means that if $u = f(x)$ is a solution to (66) and we apply v to it, this will be transformed into a new solution $u = \widetilde{f}(x)$, where we write formally $\widetilde{f} = e^{\lambda v} f$.

Thus, we can use any symmetry of $\mathcal{G}$ to obtain new solutions from known solutions; obviously, this is especially interesting when we can obtain nontrivial solutions from trivial ones. Although this could, at first sight, seem more like

[14]In the sense of a reduction valid for the whole set of solutions to the equation, as was the case for ODEs.

wishful thinking than a concrete possibility, the following (well known) example shows that this is indeed possible.

Example 4. Let us consider the heat equation in one spatial dimension, i.e. for $u = u(x,t)$, $x \in \mathbf{R}^1$; this is written as $u_t = u_{xx}$. Together with some "obvious" symmetries, corresponding to its scaling properties, and its being an autonomous linear homogeneous equation[15], this has two further symmetry vector fields, i.e.

$$\begin{aligned} v_1 &= 4t^2\partial_t + 4xt\partial_x - (x^2+2t)u\partial_u \\ v_2 &= 2t\partial_x - xu\partial_u \ . \end{aligned} \tag{67}$$

In particular, v_1 generates the transformations

$$(x,t,u) \to (\widetilde{x},\widetilde{t},\widetilde{u}) = \left(\frac{x}{D^2}, \frac{t}{D^2}, De^{-\lambda x^2/(4D^2)}u\right) \tag{68}$$

where λ is a constant, and $D = (1-\lambda t)$. In this way the trivial solution $u = f(x,t) \equiv 1$ is mapped into $\widetilde{u} = \widetilde{f}(\widetilde{x},\widetilde{t}) = De^{-\lambda D^2 \widetilde{x}^2/4}$; by writing $\nu = D^{-1}$ (and dropping the tildes) we have, in this way, the gaussian solutions

$$u(x,t) = \frac{1}{\nu}\, e^{-(\lambda x^2)/(4\nu^2)} \ . \tag{69}$$

We have thus obtained the fundamental solution from a trivial one.

In this way, once we know a special solution $u = f(x,t)$ to (66), we can generate the whole orbit of f under G (the group generated by $\mathcal{G}$), and thus obtain a G-orbit of solutions to (66).

It should be mentioned that if $G_f^0 \subseteq G$ (with algebra $\mathcal{G}_f^0 \subseteq \mathcal{G}$) is the subgroup of G which leaves f invariant, i.e. that maps f to itself, then the G-orbit of f corresponds to the action of G/G_f^0.

Let us now go on to consider the way in which knowledge of $\mathcal{G}$ can help in determining special solutions to (66), without assuming to already know a solution. The main idea is to consider ansatzes based on the symmetry, i.e. look for solutions which are invariant under subgroups of G; in the notation introduced above, this means that we fix G_f^0. If we fix a certain subgroup G_0 of G to be the G_f^0, we can then determine – again by the method of characteristics, similarly to what discussed in the previous section – a basis $\zeta_1, ..., \zeta_k$ for the ring of the invariants of G_0. Then, by hypothesis, it must be possible to write the solution in terms of these. Thus, we can rewrite (66) with this ansatz, i.e. not in terms of the (x,u) but only of the $\zeta_1,...,\zeta_k$. We then obtain a *symmetry reduction* of our initial equation.

[15] This means that the equation is invariant under translations in x and t (generated by ∂_x and ∂_t), under rescaling of u (generator $u\partial_u$) or addition to u of an arbitrary solution of the heat equation itself (generators $\beta(x,u)\partial_u$; this corresponds to the linear superposition principle), and the scalings $(x,t) \to (\lambda x, \lambda^2 t)$ generated by $x\partial_x + 2t\partial_t$.

Notice that this is different in nature from the symmetry reduction considered in the case of ODEs. In that case, there was a correspondence between the solutions of the symmetry reduced equation and the full set of solutions of the original equation. In this case, instead, the solutions of the G_0-reduced equations correspond to a subset of the solutions of the original equation, i.e. those for which $G_0 \subseteq G_f^0$.

It is quite clear that the larger the group G_0, the smaller the number k of independent invariants, and thus the easier the reduced equation is to solve; in particular, if $k = 1$, we are reduced to an ordinary differential equation (see the example below). On the other side, for large G_0, the orbit $G/G_f^0 \subseteq G/G_0$ of any given solution is smaller, i.e. we determine "fewer solutions" when we apply the group to obtain new solutions from the one we have determined.

In view of the possibility of applying the group transformations to obtain new solutions, it is clear that one does not need to consider different G_0 which are equivalent under the action of G transformations. As the solution $\widetilde{f} = g(f)$, with $g \in G$, is invariant under $G_{\widetilde{f}}^0 = gG_f^0g^{-1}$, this means that we need to consider only one subgroup for each conjugacy class of subgroups in G. A set $\mathcal{A}$ of subgroups of G such that any subgroup of G is conjugate to one and only one element of $\mathcal{A}$ (so that no two elements of $\mathcal{A}$ are conjugated) is called an *optimal set* of subgroups [262]. Determination of an optimal set is strongly related to the problem of *stratification* of M.

`Example 5`. Let us consider the three-dimensional Laplace equation, i.e.

$$u_{xx} + u_{yy} + u_{zz} \ = \ 0 \ . \tag{70}$$

This is obviously invariant under the full group $G = SO(3)$ of three dimensional rotations. Let us then fix $G_0 = G$. The only invariant of $SO(3)$ is r, and we do actually prefer to use $\zeta = r^2 = x^2 + y^2 + z^2$. With the ansatz $u = f(r^2)$, (70) reads

$$2\zeta \frac{d^2 f(\zeta)}{d\zeta^2} \ + \ 3\frac{df(\zeta)}{d\zeta} \ = \ 0 \ , \tag{71}$$

and we have indeed a reduction to an ODE.

Let us now consider $G_0 = SO(2)$ corresponding to rotations around, say, the z axis; in this case we have two invariants, e.g. $\zeta_1 = x^2 + y^2$ and $\zeta_2 = z$. Requiring that $u = f(\zeta_1, \zeta_2)$, (70) reads

$$4\zeta_1 \frac{\partial^2 f}{\partial \zeta_1^2} + \frac{\partial^2 f}{\partial \zeta_2^2} + 4\frac{\partial f}{\partial \zeta_1} \ = \ 0 \ , \tag{72}$$

and we are reduced to a PDE in two independent variables instead of three; applying the rotation group [or actually $SO(3)/SO(2)$] to any solution of (72) we will obtain a new solution that is invariant under rotations around a different axis.

In the above discussion, we have considered the ansatz $G_f^0 = G_0 \subseteq G$. However, it is possible that a solution be invariant under a vector field which is *not* a symmetry of the equation; in other words, in general $G_f \not\subseteq G$.

Example 6. For $u = u(x,y)$, $u \in \mathbf{R}$, and α, β nonzero real constants, consider the PDE

$$u_{xx} + u_{yy} - (\beta/\alpha)\, y u_y + 2\alpha u - \alpha(x^2+y^2)u - \beta y^2 u \, . \tag{73}$$

This is easily seen to be not invariant for rotations in the (x,y) plane, i.e. under $v = y\partial_x - x\partial_y$. On the other hand, the function

$$f(y) \;=\; e^{-(x^2+y^2)/2} \tag{74}$$

is obviously rotationally invariant, and gives a solution to (73).

If a vector field v is not a symmetry for the equation $F = 0$, but either it leaves invariant a specific solution $u = f(x)$ to $F = 0$, or leaves globally invariant a subset of the solutions[16] to $F = 0$, then we say it is a *conditional symmetry*. This term was introduced, together with the method briefly described below, by Levi and Winternitz [221]. Although some confusion might be possible with closely related methods (which we will not discuss here) [101,140,264,272,273,328,329], the situation was recently clarified by Pucci [274]. It should be mentioned that, according to Vorob'ev [329], the basis for the method of conditional symmetries goes back to Felix Klein.

When we have some reason to look for solutions that are invariant under a specific conditional symmetry v (e.g. on the basis of physical considerations, or of experimental observations), we can proceed essentially in the same way as before, i.e. by determining the invariants $\zeta_1, ..., \zeta_k$ of v, and using the ansatz $u = f(\zeta_1, ..., \zeta_k)$. However, in most cases we do not know a priori what the conditional symmetries should be, and we have to face the preliminary problem of determining them.

In order to do this, and actually determine at once the conditional symmetries and the solutions invariant under them, we can proceed as follows [221,342]. First of all, we write a generic vector field in M as

$$v \;=\; \sum_{i=1}^{q} \xi^i(x,u)\frac{\partial}{\partial x^i} \;+\; \sum_{\alpha=1}^{p} \varphi^\alpha(x,u)\frac{\partial}{\partial u^\alpha} \;\; ; \tag{75}$$

now, it is clear that $v \in \mathcal{G}_f$, i.e. the graph $\Gamma_f = \{(x,u) : \; u = f(x)\}$ is invariant under v (see above) if and only if

$$\varphi^\alpha \;=\; \sum_{i=1}^{q} \xi^i \frac{\partial f^\alpha}{\partial x^i} \;\; . \tag{76}$$

This means that we have a first order PDE that f has to satisfy. This is readily translated into an equation for u, i.e.

$$\mathcal{F}\left(x, u^{(1)}\right) \equiv \sum_i \left[\xi^i u_i\right] - \varphi \;=\; 0 \tag{77}$$

---

[16] I.e., solutions in the subset are exchanged under v.

(this is a vector equation, as the one for F, since u and φ are vectors in $\mathbf{R}^p$).

We can then consider that the system

$$\begin{aligned} F\left(x, u^{(m)}\right) &= 0 \\ \mathcal{F}\left(x, u^{(1)}\right) &= 0 \, , \end{aligned} \tag{78}$$

identifies the solutions to (66) which are invariant under v.[17] Notice that this method can be used both for $v \in \mathcal{G}$ and for v a conditional symmetry. Indeed, by definition, $v \notin \mathcal{G}$ is a conditional symmetry if and only if (78) admits a solution.

In looking for solutions to (78) we can still use the usual methods, and in particular we can use symmetry analysis, and write down the determining equations for the symmetry of (78). It is quite clear that v will always be, by construction, a symmetry of $\mathcal{F}\left(x, u^{(1)}\right)$.

Notice, however, that now v, i.e. ξ, φ, plays a double role: it appears not only as the symmetry vector field, but in the equation as well. Due to this, the determining equations will now be *nonlinear.*

Notice also that, as $\mathcal{F}$ is different for different vector fields, the conditional symmetries of a given equation $F = 0$ do not, in general, form an algebra. This is also clear from their definition: if two conditional symmetries are related to invariance of different solutions, there is no reason why their commutator should be a symmetry of any solution.

Finally, we would like to note that conditional symmetries, and corresponding conditionally conserved quantities, can also exist for ODEs. Actually, these were already known to Amaldi and Levi Civita [7], and have been reconsidered recently e.g. in [286,342]. While the relation among conditionally conserved quantities and invariant manifolds for the flow of the ODE is quite clear, it is not clear if the methods developed in the frame of (evolution) PDEs can be of actual use in the study or determination of invariant manifolds for ODEs.

Example 7. Consider the system of ODEs

$$\begin{aligned} \dot{x} &= h(x,y) \\ \dot{y} &= 2xh(x,y) \; . \end{aligned} \tag{79}$$

This is not – for general h – invariant under $v = 2x\partial_x + y\partial_y$, but the curve $y = x^2$ is invariant under v and under the flow of the system.

---

[17] Indeed, we can think of solving first the $\mathcal{F} = 0$. Substituting in $F = 0$ for this, we obtain the reduced equation under the ansatz that the solution be v-invariant.

5. On the Application of Symmetry Methods

To conclude this chapter, in which we have briefly summarized the basic features of symmetry analysis for differential equations (we stress again that a number of more complete introductions are available [55,146,260,262,263,305]), we would like to draw the readers attention of the to three simple but very important facts.

The first is that all the methods that we have mentioned – and, more generally, all methods of symmetry analysis for differential equations – are based on the knowledge of *some* symmetry. That is, we *do not* need to know the whole symmetry algebra $\mathcal{G}$ in order to apply these methods, and actually we can apply them even if we know merely *one* symmetry vector field (although, of course, the more we know about $\mathcal{G}$ the more effective the application of symmetry methods is).

This means that not only can we use these methods even when $\mathcal{G}$ is infinite-dimensional, and cannot be determined completely (as will usually be the case for dynamical systems, the main focus of the next chapters), but also that in general we do not have to strive to determine the complete $\mathcal{G}$, and we can start applying our methods as soon as we know some symmetry (maybe trivial) of the equation we want to study.

The second remarkable feature is that the procedures for determining the symmetry – and implement the symmetry reduction – are completely algorithmic. That is, we have a well defined procedure to follow, which requires the solution of a large but linear system of PDEs to determine the symmetries, and which yields a definite prescription for the implementation of the symmetries thus determined to obtain a symmetry reduction (in the case of ODEs) or special solutions (for PDEs) of the equation we are studying.

This also means – this is the third remark – that these procedures are amenable to implementation by computers in symbolic manipulation programs. Indeed, this is now a standard tool for symmetry analysis of differential equations, and a number of programs, using different languages, are available. For an overview of these, we refer to [34,78,180].

Appendix. The Prolongation Formula

Let us consider a function $f : X \to U$ together with its partial derivatives of order up to m. This defines a function $f^{(m)}$, called m-th prolongation of f,

$$f^{(m)} : X \to U^{(m)} , \tag{A.1}$$

which can be identified with its graph $\Gamma_{f^{(m)}}$ or, equivalently, with the manifold $\Gamma_f^{(m)}$ obtained by lifting the manifold $\Gamma_f \subset M$ to $M^{(m)}$ compatibly with the contact structure in $M^{(m)}$. We have

$$\Gamma_f^{(m)} \equiv \Gamma_{f^{(m)}} . \tag{A.2}$$

If now we have a vector field X on M, or its prolongation $X^{(m)}$ on $M^{(m)}$, this will induce an action on $f^{(m)}$, similarly to what happened for f. We will give this action in recursive form. Let us start by studying the induced action on $f^{(1)}$. From the above discussion, we had

$$\widetilde{f}^i_\varepsilon(x) = f^i(x) + \varepsilon \left[\varphi^i(x, f(x)) - \xi^j(x, f(x)) \cdot \frac{\partial f^i}{\partial x^j}(x) \right] + O(\varepsilon^2) , \tag{A.3}$$

so that now we have immediately

$$\frac{\partial \widetilde{f}^i_\varepsilon}{\partial x^h}(x) = \frac{\partial f^i}{\partial x^h} + \varepsilon \left[\frac{\partial \varphi^i}{\partial x^h} + \frac{\partial \varphi^i}{\partial u^\ell}\frac{\partial f^\ell}{\partial x^h} + \left(\frac{\partial \xi^j}{\partial x^h} + \frac{\partial \xi^j}{\partial u^\ell}\frac{\partial f^\ell}{\partial x^h} \right) \frac{\partial f^i}{\partial x^j} \right] + O(\varepsilon^2) . \tag{A.4}$$

We can obtain more compact formulas (which is quite good for higher order derivatives) by introducing the *total derivatives* D_i, which are

$$D_i \equiv \partial_i + \sum_{J,k} \frac{\partial u^k_J}{\partial x^i} \partial_{u^k_J} . \tag{A.5}$$

(Here the u, u_i, u_J should be considered to be independent, coherently with the jet space approach.)

In this way (A.5) becomes

$$\frac{\partial \widetilde{f}^i_\varepsilon}{\partial x^h} = \frac{\partial f^i}{\partial x^h} + \varepsilon \left[D_h \left(\varphi^i - \xi^j \frac{\partial f^i}{\partial x^j} \right) + \xi^j \frac{\partial^2 f^i}{\partial x^j \partial x^h} \right] + O(\varepsilon^2) \tag{A.6}$$

Since f is completely arbitrary, this describes the action induced by $X \in$ *diff*(M) on $M^{(1)}$,

$$X^{(1)} : u^i_h \to u^i_h + \varepsilon \left[D_h(\varphi^i - \xi^j u^i_j) + \xi^j u^i_{jh} \right] + O(\varepsilon^2) , \tag{A.7}$$

and if we write the first prolongation of X as

$$X^{(1)} = \xi^i \partial_i + \varphi^h \partial_{u^h} + \Phi^{(i)}_\alpha \partial_{u^\alpha_i} , \tag{A.8}$$

we have the *first prolongation formula*

$$\Phi_\alpha^{(i)} = D_i(\varphi^\alpha - \xi^h u_h^\alpha) + \xi^h u_{h,i}^\alpha \ . \tag{A.9}$$

The same approach makes it possible to obtain the general prolongation formula, which gives the coefficients $\Phi_\alpha^{(J)}$ in $X^{(m)} = X + \sum \Phi_\alpha^{(J)} \partial_{u_J^\alpha}$. In fact, $M^{(2)}$ is a subspace of $L^{(1)}$, where $L = M^{(1)}$. Using this recursively and also the fact that $X^{(n+1)}$ is the first prolongation of $X^{(n)}$, we get

$$\Phi_\alpha^{(J)} = D_J(\varphi^\alpha - \xi^h u_h^\alpha) + \xi^h u_{h,J}^\alpha \ . \tag{A.10}$$

Clearly, (A.10) is by construction compatible with the contact structure of $M^{(m)}$. It is quite obvious that $\mathcal{G}_{f^{(m)}} = \mathcal{G}_f^{(m)}$, and $\mathcal{G}_f^{(m)}$ is the prolongation of $\mathcal{G}_f \subset \mathit{diff}(M)$. Notice also that, by (A.10), $\Phi^{(J)}$ depends on x, u and derivatives of u of order up to $|J|$, i.e.

$$\Phi_\alpha^{(J)} \equiv \Phi_\alpha^{(J)}(x, u^{(|J|)}) \ . \tag{A.11}$$

Thus, the $X^{(m)}$ are actually legitimate vector fields on $M^{(m)}$.

In Sect. 2 of this chapter, see (20), it was stated that the algebra $\mathcal{G}$ generated by $\{X_1, ..., X_n\}$ and the prolonged algebra $\mathcal{G}^{(m)}$ generated by $\{Z_1, ..., Z_n\}$ – with $Z_i = X_i^{(m)}$ – are isomorphic, and

$$\left[X_i^{(m)}, X_j^{(m)}\right] = [X_i, X_j]^{(m)} \ . \tag{A.12}$$

This can be proved by direct computation, or using properties of Lie brackets [262]. We give a simple proof here. Let

$$\begin{array}{lll} X_1 = \xi\partial_x + \varphi\partial_u & ; & X_1^{(1)} = X_1 + \Phi\partial_{u_x} \\ X_2 = \eta\partial_x + \psi\partial_u & ; & X_2^{(1)} = X_2 + \Psi\partial_{u_x} \end{array} \ ; \tag{A.13}$$

we also write

$$Y = [X_1, X_2] = \vartheta\partial_x + \rho\partial_u \ . \tag{A.14}$$

As for $\left[X_1^{(1)}, X_2^{(1)}\right]$, it is easy to see that, since we are supposing that ξ, φ, η, ψ depend only on (x, u) and not on u_x, we can write it in the form

$$\left[X_1^{(1)}, X_2^{(1)}\right] = \vartheta\partial_x + \rho\partial_u + \chi\partial_{u_x} = Y + \chi\partial_{u_x} \ . \tag{A.15}$$

But $X_1^{(1)}, X_2^{(1)}$ belong to the algebra $\mathit{diff}^{(1)}(M)$, so their commutator belongs to $\mathit{diff}^{(1)}(M)$ as well, i.e. χ must satisfy the prolongation formula

$$\chi = D_x(\rho - \vartheta u_x) + \vartheta u_{xx} \tag{A.16}$$

and we have

$$\left[X_1^{(1)}, X_2^{(1)}\right] = Y^{(1)} \ . \tag{A.17}$$

This extends immediately to m-th prolongations, as in (5):

$$\left[X_1^{(m)}, X_2^{(m)}\right] = Y^{(m)} \ . \tag{A.18}$$

We only have to notice that the prolongation operator is linear, so that $pr^{(m)}(X_1 + X_2) = X_1^{(m)} + X_2^{(m)}$, to conclude that the algebras generated by the X_i's and the one generated by the $X_i^{(m)}$ are isomorphic.

II. Dynamical Systems

The purpose of this chapter is not to provide an introduction to dynamical systems, but only to recall some important concepts and to define notations. We will closely follow the conventions of [11,12].

Here, by *dynamical system* (DS in the following) on a smooth manifold n-dimensional M we mean a smooth tangent vector field X_f defined on M. We will consider, for ease of discussion, M as embedded in $\mathbf{R}^m$, and consider coordinates $x^1, ..., x^m$ in $\mathbf{R}^m$. We then have a DS in the form of a system of autonomous first order ODEs for the x as

$$\dot{x} = f(x) \quad ; \quad x \in \mathbf{R}^m \ , \ f : M \to \mathrm{T}M \ . \tag{1}$$

Correspondingly, we express the vector field X_f in x coordinates as

$$X_f = f^i(x) \frac{\partial}{\partial x^i} \equiv f(x) \partial_x \ . \tag{2}$$

In the following, we will take for granted that f is a smooth ($\mathcal{C}^\infty$, or even analytical) function, and correspondingly X_f is a smooth vector field (VF).

We will denote by $\mathcal{M}$ the set of smooth vector fields on M, by $\mathcal{F}$ the set of smooth scalar (real) functions $f : M \to \mathbf{R}$, and by $\mathcal{V}$ the set – isomorphic to $\mathcal{M}$ – of smooth vector (real) functions $f : M \to \mathrm{T}M$.

1. Dynamical Systems and Flows

To the VF X_f we associate a direction field $X_{\widehat{f}}$, defined on $M_f = M \backslash [f^{-1}(0)]$ as

$$X_{\widehat{f}} = \frac{f^i(x)}{|f(x)|} \frac{\partial}{\partial x^i} \ . \tag{3}$$

Here $|f(x)|$ is the standard $\mathbf{R}^m$ modulus of $f(x)$, i.e. $|f(x)| = \left(\sum_{i=1}^m [f^i(x)]^2 \right)^{(1/2)}$, and $f^{-1}(0) \subseteq M$ is the set of points $x \in M$ on which $f(x) = 0$.

Let us now consider a neighborhood $\mathcal{B}$ of some point $x_0 \in \mathcal{B} \subset M$. If $X_{\widehat{f}}$ identifies a parallel field of direction (in a neighborhood $\mathcal{B} \subset M$), we say it is a rectilinear field (in the neighborhood $\mathcal{B} \subset M$). It is well known [11] that:

Theorem 1 (Newton). *Every smooth direction field $X_{\widehat{f}}$ is rectifiable in a neighborhood of any point in M_f. If $\widehat{f}$ is of class C^r on M_f, then the rectifying diffeomorphism can also taken to be of class C^r.*

Corollary 1. *Let $X_f = f(x)\partial_x$ be a vector field on M, of class C^r. For any point x_0 such that $f(x_0) \neq 0$, there exists a neighborhood $\mathcal{B} \subset M$ of x_0*

and a C^r diffeomorphism $y = \Psi(x)$ defined on $\mathcal{B}$ which reduces the vector field $X_f = f^i(x)\partial_i$ to the local form $X_f = (\partial/\partial y^1)$.

The rectifying diffeomorphism Ψ can be explicitly constructed, and its construction is actually equivalent to the Newton method of integrating differential equations by series [11].

Theorem 1 and Corollary 1 tell us that if we are interested in the local behavior of differential equations, no "interesting" behavior can arise if we are not in the neighborhood of a point $x_0 \in f^{-1}(0)$. Thus, the local analysis of ODEs will naturally focus on points x_0 such that $f(x_0) = 0$.

In most of the following chapters, we will be concerned with a perturbative setting in the neighborhood of a fixed point x_0. This will mean that we are restricted, essentially, to a local chart and so to the tangent space in x_0 to the manifold M, $\mathrm{T}_{x_0}M \simeq \mathbf{R}^n$. Thus, we can just assume, with no loss of generality, that $M = \mathbf{R}^n$. Keeping to the general notation, however, does not present any additional difficulty, and will help in clarifying some geometrical structure.

The vector field X_f defines a one-parameter semigroup of diffeomorphisms of M, $\Phi_f(t;x) : \mathbf{R}_+ \times M \to M$, which represents the time t evolution of solutions issuing from x at time $t = 0$:

$$\frac{\mathrm{d}\Phi_f(t;x)}{\mathrm{d}t}\Big|_{t=0} = f(x) \ . \tag{4}$$

The $\Phi_f(t;x)$ is also called the *flow* of X_f (or, with abuse of notation, of f). Notice that this is, in general, only a semigroup[1] on M, and a group on M_f.

From Theorem 1 it also follows that [11]:

Corollary 2. *The flow $\Phi_f(t;x)$ of the C^r vector field X_f is defined, continuous and smooth of class C^r in the neighborhood of each point $x \in M$.*

The set of points $\Phi_f(t;x)$ is called an *integral curve* for X_f, and represents a solution of the ODE (2). Corollary 2 implies then the

Theorem 2. *For each point $x \in M_f$, there is at least one integral curve which passes through x.*

Theorem 3. *If two integral curves have a common point $x \in M_f$, then they coincide in a neighborhood of x_0.*

Theorem 4. *The flow $\Phi_f(t;x)$ depends smoothly on x.*

The above theorems are the theorems of (local) existence, prolongability, uniqueness, and differentiability with respect to initial conditions, for solutions of a first order ODE.

If we have two vector fields $X_f, X_g \in \mathcal{M}$, and correspondingly two flows Φ_f, Φ_g, we can compare the two, and introduce several notions of equivalence.

---

[1] In general, $\Phi_f(t;x)$ is a local group in t; i.e. for t small enough we have indeed a group, while for larger t we have only a semigroup.

Definition. *The flows Φ_f and Φ_g are conjugate (or equivalent) if a one-to-one mapping $\Lambda : M \to M$ exists such that, for any $x \in M$ and all $t \in \mathbf{R}_+$, $\Lambda(\Phi_f(t; x)) = \Phi_g(t; \Lambda(x))$.*

If the map Λ can be chosen to be a linear isomorphism, then we say the flows are linearly equivalent, or linearly conjugate; if the map Λ can be chosen to be a diffeomorphism, we say the flows are differentiably equivalent, or differentiably conjugate; finally, if the map Λ can be chosen to be a homeomorphism, we say that the flows are topologically equivalent, or topologically conjugate.[2]

It is obvious that any flow is conjugate to itself (just consider Λ to be the identity on M); however, the determination of all the mapping Λ through which a flow is self-conjugate is, in general, not an easy task. A map $\Lambda : M \to M$ that realizes a conjugation of the flow Φ_f with itself is also called an *intertwining map* for X_f. Clearly, the set of intertwining maps for a given X_f form a group. We can also consider a map realizing a linear (respectively, differential or topological) conjugation of the flow with itself. Such linearly (respectively, differentiably or topologically) intertwining maps also form a group. In the next chapter, smooth intertwining maps will be seen as symmetries of the vector field, and studied in detail.

Exercise 1. Prove that linear conjugacy implies differential conjugacy, and that the latter implies topological conjugacy.

Exercise 2. Consider the vector fields in $\mathbf{R}^1$ $X = x\partial_x$ and $Y = 2x\partial_x$. Are these linearly conjugate ?

Exercise 3. Prove that the relations of linear, differential, and topological equivalence are indeed equivalence relations.

Exercise 4. Prove that for a given flow Φ_f, the sets of intertwining, linearly intertwining, differentiably intertwining, and topologically intertwining maps form indeed a group, as stated above.

Exercise 5. Characterize the set of linearly intertwining maps for a generic vector field $X_f = f^i(x)\partial_i$, and for a linear vector field $X_A = (A_{ij}x^j)\partial_i$.

2. Singular Points and Invariant Manifolds

As mentioned in the previous section, it is natural to focus our attention on the singular points for the vector field X_f, i.e. on the points in $f^{-1}(0) \subset M$. From now on, we fix such a point x_0, and we have therefore $f(x_0) = 0$. We also denote by E the tangent space to M in x_0, i.e. $E = \mathrm{T}_{x_0}M$.

We can then consider the linearization of f at x_0; this will be an $(n \times n)$ matrix $A = (Df)(x_0)$. To this are associated the eigenvalues $\lambda_1, ..., \lambda_m$ and the corresponding eigenvectors $v^{(1)}, ..., v^{(n)}$, with $Av^{(i)} = \lambda_i v^{(i)}$.

---

[2] Exercises 1–5, concerning conjugacy of flows, are taken from [11].

In the same way as we considered equivalence among flows on M, we can consider equivalence among flows in a neighborhood of a fixed point x_0. In particular, a first analysis of the behavior of X_f – that is, of Φ_f – in the vicinity of x_0 is provided by considering its linearization X_{Ax}, denoted for short by X_A, and the corresponding linear flow Φ_A. Equivalence of linear flows is described by the following lemmata.

Lemma 1. *Let A, B be $(n \times n)$ matrices having simple eigenvalues. Then Φ_A and Φ_B are linearly equivalent if and only if all the eigenvalues of A and B coincide.*

Lemma 2. *Let A, B be $(n \times n)$ matrices having simple eigenvalues. Then Φ_A and Φ_B are differentiably equivalent if and only if they are linearly equivalent.*

Lemma 3. *Let A, B be $(n \times n)$ matrices having only eigenvalues with nonzero real parts. Then Φ_A and Φ_B are topologically equivalent if and only if A and B have the same number of eigenvalues with negative real parts (and therefore of eigenvalues with positive real parts as well).*

A fixed point x_0 such that $A = (Df)(x_0)$ has no eigenvalue with zero real part is called a *hyperbolic fixed point* for X_f.

We say that X_f is locally equivalent to X_g at a common fixed point x if there is a neighborhood $\mathcal{B} \in M$ containing x_0 on which the flows Φ_f and Φ_g are equivalent.

The celebrated Hartmann-Grobmann theorem states local equivalence of a vector field and its linearization at a fixed point:

Theorem 5 (Hartmann-Grobmann). *Let x_0 be a fixed point for X_f, and let X_A be the linearization of X_f at x_0, $A = (Df)(x_0)$. If x_0 is a hyperbolic fixed point, then Φ_f is locally topologically equivalent to Φ_A.*

Corollary (Topological classification theorem). *Let X_f, X_g be vector fields having a common fixed point at x_0, and let $A = (Df)(x_0)$, $B = (Dg)(x_0)$. Let A, B have no eigenvalue with zero real part. Then, X_f and X_g are locally topologically equivalent at x_0 if and only if A and B have the same number of eigenvalues with positive (and therefore also with negative) real parts.*

We can divide the finite set of eigenvalues of $A = (Df)(x_0)$ into three groups: the eigenvalues with negative, zero, and positive real part. Correspondingly, the eigenvectors relative to eigenvalues with negative real part will span a linear space $E_- \subseteq E = \mathrm{T}_{x_0}M$; the eigenvectors relative to eigenvalues with zero real part will span a linear space $E_0 \subseteq E = \mathrm{T}_{x_0}M$; and the eigenvectors relative to eigenvalues with positive real part will span a linear space $E_+ \subseteq E = \mathrm{T}_{x_0}M$.

Denoting the number of eigenvalues in each group (counted with their geometric multiplicities) by, respectively, n_-, n_0, n_+, we have that $E_- \simeq \mathbf{R}^{n_-}$, $E_0 \simeq \mathbf{R}^{n_0}$, and $E_- \simeq \mathbf{R}^{n_+}$.

The space E_- is also called the (linear) *stable space* at x_0; the space E_+ is called the (linear) *unstable space* at x_0; and the space E_0 is called the (linear) *center space* at x_0. Notice that these are generalized eigenspaces. Clearly,

$$\begin{array}{lll} \lim_{t\to+\infty} & \Phi_A(t;x) = x_0 & \forall x \in E_- \ , \\ \lim_{t\to-\infty} & \Phi_A(t;x) = x_0 & \forall x \in E_+ \ . \end{array} \tag{5}$$

When we consider the full (nonlinear) flow Φ_f in a neighborhood $\mathcal{B}$ of x_0, there are *invariant local manifolds* W_-, W_+, W_0 (submanifolds of $\mathcal{B}$) which are tangent in x_0 to, respectively, E_-, E_+, E_0. The invariance alluded to is the one under the flow Φ_f; that is, if $x \in W_i \subset \mathcal{B}$ (with $i = +, -, 0$), then $X_f(x) \in \mathrm{T}_x W_i$. These manifolds will be called, respectively, the (local) *stable manifold* (W_-), *unstable manifold* (W_+), and *center manifold* (W_0).

More precisely, let us first consider the case x_0 is a hyperbolic fixed point, so that $E = E_+ \oplus E_-$. Then we have [16,231,279]

Theorem 6 (Hadamard-Perron). *Let X_f be a C^r (respectively, C^∞ or analytic) vector field on M, x_0 be a hyperbolic fixed point for X_f, and let $E_- \subseteq \mathrm{T}_{x_0}$, $E_+ \subseteq \mathrm{T}_{x_0}M$ be the stable and unstable linear spaces for $A = (Df)(x_0)$. Then there are C^r (respectively, C^∞ or analytic) submanifolds $W_- \subseteq M$, $W_+ \subseteq M$ such that $X_f : W_\pm \to \mathrm{T}W_\pm$, and such that $\mathrm{T}_{x_0}W_\pm = \mathrm{T}_{x_0}E_\pm$. Moreover, for $x \in W_\pm$, $\Phi_f(t;x)$ tends exponentially to x_0 for $t \to \mp\infty$.*

Note that if x_0 is hyperbolic, then $W_\pm$ are *normally hyperbolic* in x_0. Moreover, let $U \subset M$ be a neighborhood of x_0, and consider the linearization $A(x) = (Df)(x)$ of f at points $x \in W_\pm \cap U$. From the smoothness of f and M, it follows that the eigenvalues $\lambda_i(x)$ of $A(x)$ are also smooth functions of x. Thus, if U is sufficiently small, $\mathrm{Re}\{\lambda_i\} \neq 0$ for all points $x \in U$ and for all $i = 1, ..., n$. Thus, we have

Lemma 4. *If x_0 is a hyperbolic fixed point for X_f, then $W_\pm$ are locally normally hyperbolic; i.e., there is a neighborhood U of x_0 such that $W_\pm \cap U$ are normally hyperbolic manifolds.*

The case of x_0 non-hyperbolic is more delicate. In particular, the center manifold is in this case not unique, and the center manifolds are in general less smooth than the vector field. We have [16,173]:

Theorem 7 (Center manifold theorem). *Let X_f be a C^r (respectively, C^∞ or analytic) vector field on M. Let x_0 be a fixed point for X_f, and let $E_- \subseteq E \equiv \mathrm{T}_{x_0}$, $E_+ \subseteq E$ and $E_0 \subseteq E$ be the stable, unstable and center linear spaces for $A = (Df)(x_0)$. Then there are C^r (respectively, C^∞ or analytic) submanifolds $W_- \subseteq M$ and $W_+ \subseteq M$, and C^{r-1} (respectively, finitely smooth) submanifolds $W_0 \subseteq M$ such that, for $\alpha = \pm, 0$, $X_f : W_\alpha \to \mathrm{T}W_\alpha$, and such that $\mathrm{T}_{x_0}W_\alpha = \mathrm{T}_{x_0}E_\alpha$. Moreover, for $x \in W_\pm$, $\Phi_f(t;x)$ tends exponentially to x_0 for $t \to \mp\infty$. The manifolds $W_\pm$ are unique, but W_0 need not be.*

In this case, the flow has two other important invariant manifolds, namely the *center-stable* and the *center-unstable* manifolds [279] that are of special importance in *bifurcation theory* [173,190,279].

Note that, as stated in the theorem, the center manifolds are, in general, only finitely smooth even for analytic vector fields. Note also that, while the local behavior (near x_0) of the flow on the stable and unstable manifolds is determined by the linear terms alone, for the flow on the center manifold we have to take into account the nonlinear terms as well. Thus, if we are interested in topological conjugacy of the flow in a neighborhood of x_0 (as in the Hartmann-Grobmann theorem), we cannot reduce the center manifold flow to a standard form. We have [16]:

Theorem 8. *Let X_f be a C^r vector field on M, with $r \geq 2$. Let x_0 be a fixed point for X_f, and let $E_- \subseteq E \equiv \mathrm{T}_{x_0}$, $E_+ \subseteq E$ and $E_0 \subseteq E$ be the stable, unstable and center linear spaces for $A = (Df)(x_0)$. Let W_+, W_-, W_0 be the stable, unstable and center manifolds, and let (x, y, w) be local coordinates such that $W_- = (x, 0, 0)$, $W_+ = (0, y, 0)$, $W_0 = (0, 0, w)$. Then, the flow Φ_f in a neighborhood U of x_0 is locally topologically conjugate to the flow of the vector field X_s which is the direct product of vector fields X_-, X_+ and X_h, where X_h is defined on W_0, where $h(w) = f(w)$, and $X_- = -x^i(\partial/\partial x_i)$, $X_+ = y^i(\partial/\partial y_i)$.*

This theorem means that, if we are interested in the topological classification, we should only consider the flow on the center manifold, i.e. the problem of "distinguishing a center and a focus". It should be stressed that the theorem concerns topological, but not differentiable, conjugacy.

It should be stressed that the theorems on invariant manifolds presented here admit generalizations in at least two important directions: First of all, we can consider a Banach space B rather than $\mathbf{R}^n$, such a generalization being relevant in the study of evolution PDEs. Second, we can consider families of vector fields depending smoothly on a *control parameter*, this being the case studied in *bifurcation theory*. The relevant extensions are discussed and studied in [279].

Exercise 6. Consider the vector fields $X = [-x + (1+\pi)y]\partial_x + [2x + (2+\pi)]y$ and $Y = (x + x^2 y)\partial_x + [(\pi + 2x^3 y)y]\partial_y$ on $M = \mathbf{R}^2$. Are these conjugate ? Are these linearly, differentiably, topologically conjugate ?

Exercise 7. Consider the same vector fields as in Exercise 6; they have a common fixed point at the origin, $x_0 = 0$. Derive the matrices describing their linearizations at x_0, $A = (Df)(0)$, $B = (Dg)(0)$. Are X_f, X_g locally topologically conjugate at x_0 ? Describe the linear stable, unstable and center spaces at x_0, and the local stable, unstable and center manifolds.

Exercise 8. Consider the vector field $X = -x^2\partial_x - y\partial_y$, which admits $x_0 = 0$ as a fixed point; determine the center manifold(s) W_0 at x_0. Is W_0 unique ? What if we require an analytic center manifold ?

3. Conserved Quantities

Let us now turn our attention toward *conserved quantities* under the flow Φ_f; these are also called *constants of motion* for X_f.

We say that[3] $\rho \in \mathcal{F}$ is a constant of motion – or a conserved quantity – for X_f if and only if

$$X_f(\rho) = 0 \quad \forall x \in M \ . \tag{6}$$

This also means that ρ is constant along all of the integral curves for the flow Φ_f.

Together with proper, or global, conserved quantities, we may sometimes wish to consider quantities which are conserved only along some of the integral curves of the flow [7,286].

We say that $\rho \in \mathcal{F}$ is a conditional constant of motion – or a conditionally conserved quantity – for X_f if and only if there is a smooth submanifold $M_\rho \subset M$ (of dimension at least one), called the support manifold, invariant under X_f and such that

$$X_f(\rho) = 0 \quad \forall x \in M_\rho \ . \tag{7}$$

This also means that ρ is constant along all of the integral curves for the flow Φ_f which go through M_ρ. Notice that the invariance of M_ρ under the flow implies that it is the union of integral curves.

One sees immediately that the set of constants of motion (or first integrals) for a given vector field X_f is a ring, and in particular is an algebra. This is clearly true both for the set $\mathcal{I}_f$ of the analytic constants of motion and for the set $\mathcal{I}_f^*$ of the meromorphic (i.e., fractional) constants of motion. It should be stressed that the set of conditional constants of motion does *not* have such a structure, as the submanifolds M_ρ corresponding to different functions are, in general, different. However, the conditional constants of motion sharing the same support manifold M_0 will have the structure of a ring and of an additive algebra.

Identifying the set of constants of motion for a given X_f requires solving a characteristic PDE. Indeed, when we write (6) using the representation of X_f in the x coordinates, we get the linear, homogeneous first order PDE

$$\sum_{i=1}^{m} f^i(x) \frac{\partial \rho}{\partial x^i} = 0 \ . \tag{8}$$

As is well known, the solution to this can be obtained by the associated characteristic equation

$$\frac{\mathrm{d}x^1}{f^1(x)} = \frac{\mathrm{d}x^2}{f^2(x)} = \dots = \frac{\mathrm{d}x^m}{f^m(x)} \ . \tag{9}$$

The solutions of this characteristic equation (9) are independent functions $\zeta_1(x), \zeta_r(x)$, that are constant along the integral curves of X_f. Thus any $\rho = \rho(\zeta)$ will be a constant of motion, and conversely any constant of motion can be written as a function of the ζ.

---

[3] Recall that $\rho \in \mathcal{F}$ means that ρ is a scalar function defined on M.

Exercise 9. Let M be the two dimensional torus $T^2 = S^1 \times S^1$, with coordinates ϑ_1, ϑ_2. Let the vector field X be expressed in these coordinates as $X = \partial_1 + \alpha\partial_2$, with α a real number. Discuss the existence of conserved quantities and of conditionally conserved quantities depending on the value of α.

Exercise 10. Let M now be a "fat torus", $M = S^1 \times S^1 \times I$ with $I = [0,1]$ the unit interval. Take coordinates (x_1, x_2, x_3) on M, with $x_1, x_2 \in S^1$ and $x_3 \in I$. Consider the vector field described in these coordinates as $X = \partial_1 + f(x_3)\partial_2$, with $f : I \to \mathbf{R}$ a smooth function. Discuss the existence of constants of motion – and of conditional constants of motion – for X, depending on f.

Exercise 11. How is your discussion in the Exercise 10 related to the property of integral curves for X of being proper submanifolds of M ?

Exercise 12. Consider the dynamical system in $\mathbf{R}^2$ given by

$$\dot{x} = -\,x^2$$
$$\dot{y} = -\,y$$

(see Exercise 8). Determine the constants of motion, and the conditional constants of motion.

4. Perturbative Expansion

As we are interested in the flow of X_f in the neighborhood of the fixed point x_0 (we can assume $x_0 = 0$ with no loss of generality and with some notational simplification; thus we will do it), it is natural to consider perturbative expansions of functions around x_0. We will consider the sets (rings) $\mathcal{V}$ and $\mathcal{F}$ of, respectively, vector and scalar functions. We will actually consider the sets (rings) of vector and scalar (possibly formal) power series $\mathcal{V}^p$, $\mathcal{F}^p$.

For our purposes, we can actually only consider functions which vanish in x_0. The set of such functions or power series, vector or scalar, smooth or analytic or only formal, do still have the structure of a ring, as is easily checked.

The sets $\mathcal{V}^p, \mathcal{F}^p$ are naturally graded by the degree of homogeneity of the polynomials. Correspondingly, they are split as

$$\mathcal{V}^p = \bigoplus_{k=0}^{\infty} \mathcal{V}_k \quad , \quad \mathcal{F}^p = \bigoplus_{k=0}^{\infty} \mathcal{F}_k \; . \tag{10}$$

Considering functions vanishing in $x_0 = 0$ then amounts to considering (as we will do in the following)

$$\bar{\mathcal{V}} := \mathcal{V}^p \backslash \mathcal{V}_0 \quad , \quad \bar{\mathcal{F}} := \mathcal{F}^p \backslash \mathcal{F}_0 \; . \tag{11}$$

We will only consider $\bar{\mathcal{V}}, \bar{\mathcal{F}}$, and drop the overbar for ease of notation.

It should be stressed, however, that when restricting to analytic functions, we are not considering any function that vanishes to all orders in perturbation around $x_0 = 0$, such as $h(x) = e^{-1/x^2}$. Also, functions f, g which differ by such a "perturbatively vanishing" function cannot be distinguished perturbatively. Exercises 8 and 12 above show that this should be carefully taken into account also when analyzing the flow in a small neighborhood of $x_0 = 0$.

As standard notation, we will expand functions in homogeneous terms, and denote by a subscript the degree of homogeneity of the terms in the expansion. Thus, for $f \in \mathcal{V}(\equiv \bar{\mathcal{V}})$ we will write

$$f(x) = \sum_{k=1}^{\infty} F_k(x) \quad , \quad F_k \in \mathcal{V}_k \ . \tag{12}$$

Similarly, for $\rho \in \mathcal{F}(\equiv \bar{\mathcal{F}})$ we write

$$\rho(x) = \sum_{k=1}^{\infty} \rho_k(x) \quad , \quad \rho_k \in \mathcal{F}_k \ . \tag{13}$$

The different objects of interest in the study of the (local) flow of X_f, such as invariant manifolds, constants of motion, and commuting flows, can then be computed perturbatively. Below we briefly discuss the resulting equations.

With the notation (12) and (13) we get for the condition (6) of $\rho \in \mathcal{I}_f$, after rearranging terms in the sum,

$$\sum_{k=1}^{\infty} \left[\sum_{m=1}^{k} \left(f_m(x) \cdot \nabla \right) \rho_{k-m} \right] = 0 \ . \tag{14}$$

Actually, the term in the square brackets has to vanish for each k, and thus we have a hierarchy of equations

$$\sum_{m=1}^{k} \left(f_m(x) \cdot \nabla \right) \rho_{k-m} = 0 \ . \tag{15}$$

We can solve these sequentially for $k = 1, 2, ...$, and in this way at each stage k the ρ_s with $s < k$ can be seen as known functions, and we are reduced to considering a linear non-homogeneous PDE for ρ_k, i.e.

$$\sum_{i=1}^{m} A_{ij} x^j \frac{\partial \rho_k}{\partial x^i} = R_k(x) \ , \tag{16}$$

where we have used $f_1(x) = Ax$, and $R_k \in \mathcal{F}_k$ is given explicitly by

$$R_k(x) = - \sum_{m=1}^{k-1} \left(f_m(x) \cdot \nabla \right) \rho_{k-m}(x) \ . \tag{17}$$

This is equivalent to requiring ρ_k to have an assigned rate of variation – that is, R_k – along the flow of the linear vector field X_A corresponding to the linearization of X_f at $x_0 = 0$.

Notice now that $\mathcal{F}_k$ is a finite-dimensional vector space, in which we can take the basis of homogeneous monomials of degree k, which we denote in multiindex notation as

$$|\mu>=x^{\mu}=x_1^{\mu_1}...x_n^{\mu_n} \ , \tag{18}$$

where $|\mu|=k$.

We then have $R(x)=r_{\mu}^{(k)}|\mu>$, and we write $\rho_k(x)=\alpha_{\mu}^{(k)}|\mu>$ (dropping the index k in intermediate computations). Equation (16) then reduces to an algebraic equation for the α_{μ}.

We denote by e_i the multiindex which has components $m_j=\delta_{ij}$, and by χ_i the function on multiindices such that $\chi_i(\mu)=0$ for $\mu_i=0$, and $\chi_i(\mu)=1$ if $\mu_i\neq 0$. Then, (16) results in

$$\sum_{|\mu|=k}\sum_{i,j} A_{ij}\ \chi_i(\mu)\ \alpha_{\mu}\ |\mu-e_i+e_j>=\sum_{|\mu|=k} r_{\mu}|\mu> \tag{19}$$

or, after a change in indices and in matrix form,

$$T_{\sigma\mu}^{(k)}\alpha_{\mu}^{(k)}=r_{\sigma}^{(k)} \ , \tag{20}$$

where the matrix T is defined as

$$T_{\sigma\mu}^{(k)}=\sum_{i,j}\chi_i(\mu)A_{ij}\delta_{(\sigma,\mu-e_i+e_j)} \ . \tag{21}$$

Thus, the perturbative construction of constants of motion is reduced to the problem of inverting the matrices $T^{(k)}$. Notice that we have, at each stage, to solve a problem that has a solution only for $r_{\mu}^{(k)}\in \mathrm{Ran}\left(T^{(k)}\right)$.

The perturbative construction of invariant manifolds goes along similar lines. We denote by $\xi_1,...,\xi_n$ the eigenvectors of A and, assuming[4] that these span $\mathbf{R}^n$, we go on to consider these as coordinates in a neighborhood $\mathcal{B}$ of $x_0=0$. Now the ξ_i are obviously constants of motion for the linearized vector fields $X_{(A-\lambda_i I)x}$, and the linear spaces E_i are identified as common zero-level sets for a certain set of the constants of motion ξ_j. In the same way, the stable and unstable invariant manifolds $W_{\pm}$ (see Sect. 2) can be seen as the common zero-level set for the same set of constants of motion under the full nonlinear X_f. Thus the perturbative construction of constants of motion, specialized to the case $\rho_1=\xi_i$, also permits us to construct perturbatively the invariant manifolds.

Similar considerations apply to the invariant center manifolds W_0 (also defined in Sect. 2 of this chapter); here the non-uniqueness of W_0 requires some care, but despite this non-uniqueness, the perturbative expansion for W_0 is well defined, and we can proceed in the same way [173,190].

Exercise 13. Consider the dynamical system in $\mathbf{R}^3$, corresponding to X_f,

[4] This is the case if A is semisimple. We could actually proceed without this assumption, by considering generalized eigenvalues and eigenspaces.

$$\begin{aligned}\dot{x} &= x - x^2 y - x^3 \\ \dot{y} &= y + x y^2 + x^3 \\ \dot{z} &= yz + x^3 z + x y^2 z \ .\end{aligned}$$

Determine perturbatively its constants of motion up to order $k = 3$.

Exercise 14. For the same dynamical system as in Exercise 13, determine the linearization $A = (Df)(0)$ of f at $x_0 = 0$, and the eigenvalues and eigenvectors of A. Consider then the matrices $B^i = (A - \lambda_i I)$. Determine the constants of motion for X_{B^i} of order up to $k = 3$. Is there any relation between these and the result of Exercise 13 ?

Exercise 15. Consider the dynamical system in $\mathbf{R}^2$ given by

$$\begin{aligned}\dot{x} &= -\, x^2 \\ \dot{y} &= -\, y\end{aligned}$$

(see Exercises 8 and 12). Determine perturbatively, up to some order N (of your choice) the constants of motion, and the conditional constants of motion.

5. Poincaré-Dulac Normal Forms

In the previous Sect. 4, we have introduced the expansion (12) in homogeneous terms for f, defining the vector field $X \equiv X_f$ or equivalently the dynamical system $\dot{x} = f(x)$.

If now we consider a change of coordinates of the form

$$y = x + h_k(x) \tag{22}$$

where $h_k \in \mathcal{V}_k$ $(k \geq 1)$, we get

$$\dot{y}^i = \dot{x}^i + \Gamma^i_j \dot{x}^j = \sum_m f_m\left(x + h_k(x)\right) \ . \tag{23}$$

Here, obviously,

$$\Gamma^i_j = \frac{\partial h^i_k}{\partial x^j} \ . \tag{24}$$

A change of coordinates of the form (22) is called a *Poincaré transformation* (or P transformation for short), and the function h_k is also called the *generator* of the P transformation. In the following we will drop the subscript k when this cannot generate any confusion.

Notice that (22) is, for small x, a near-identity transformation, thus, it is surely invertible in a small enough neighborhood of the origin. Therefore, we can apply

$$\Lambda = (I + \Gamma)^{-1} \tag{25}$$

on (6), and get the P transformed dynamical system in the form

$$\dot{x} = \widetilde{f}(x) \equiv \Lambda \sum_{m=0}^{\infty} f_m \left(x + h_k(x)\right) \ . \tag{26}$$

In order to have this again as a series in homogeneous terms, i.e. in order to identify the $\widetilde{f}_m$, we should consider power series expansions for Λ and for $f_m(x + h(x))$.

Introducing multiindex notation $\mathcal{J} = (j_1, ... j_n)$, so that $|\mathcal{J}| = \sum_i j_i$,

$$\partial_{\mathcal{J}} \equiv \frac{\partial^{|\mathcal{J}|}}{\partial x_1^{j_1} ... \partial x_n^{j_n}} \ , \tag{27}$$

and $h^{\mathcal{J}} = (h^1)^{j_1} ... (h^n)^{j_n}$ (here h^i is the i-th component of the vector h), we define the operators

$$\Phi_h^r = \frac{1}{r!} \sum_{|\mathcal{J}|=r} h^{\mathcal{J}} \partial_{\mathcal{J}} \ . \tag{28}$$

With these, we have

$$f\left(x + h(x)\right) \ = \ \sum_{r=0}^{\infty} \Phi_h^r \ f(x) \ . \tag{29}$$

Notice that

$$h \in \mathcal{V}_k \quad \Rightarrow \quad \Phi_h^r : \mathcal{V}_m \to \mathcal{V}_{m+rk} \ . \tag{30}$$

As for Λ, we just recall that

$$(I + \varepsilon M)^{-1} = \sum_{k=0}^{\infty} (-\varepsilon)^k M^k \ , \tag{31}$$

and therefore

$$\Lambda = \sum_{k=0}^{\infty} (-1)^k \Gamma^k \ , \tag{32}$$

Finally, our transformed system is

$$\dot{x} \ = \ \sum_{m=0}^{\infty} \sum_{r=0}^{\infty} \sum_{s=0}^{r} \left[(-1)^s \ \Gamma^s \ \Phi_h^{r-s} \right] \ f_m(x) \ . \tag{33}$$

In other words, in the P transformation with generator $h_k \in \mathcal{V}_k$ each term f_m is transformed into a new $\widetilde{f}_m$ given (with $[x]$ the integer part of x) by

$$\widetilde{f}_m \ = \ f_m \ + \ \sum_{p=1}^{[m/k]} \left[\sum_{s=0}^{p} (-1)^s \Gamma^s \Phi_{h_k}^{p-s} \right] f_{m-kp} \ . \tag{34}$$

From this formula it follows, in particular, that – as is well known – the terms of degree smaller than k are not changed at all,

$$\widetilde{f}_m = f_m \quad \forall m < k \ , \tag{35}$$

and the term of degree k is changed according to

$$\widetilde{f}_k = f_k + [\Phi_h - \Gamma] f_0 . \tag{36}$$

We also notice that, similarly, for $\nu < k$, the term of degree $k + \nu$ is changed into[5]

$$\widetilde{f}_{k+\nu} = f_{k+\nu} + [\Phi_h - \Gamma] f_\nu \quad (0 < \nu < k) . \tag{37}$$

We will use the antisymmetric bracket $\{.,.\} : \mathcal{V} \times \mathcal{V} \to \mathcal{V}$, defined as

$$\{f, g\} \ := \ (f \cdot \nabla) g \ - (g \cdot \nabla) f . \tag{38}$$

Together with this, we define the operators

$$L_k = \{f_k, .\} . \tag{39}$$

Note that

$$L_k \ : \ \mathcal{V}_m \ \to \ \mathcal{V}_{m+k} . \tag{40}$$

Recalling that $\Gamma^i_j = \nabla_j h^i$ and that $\Phi_h = h^i \nabla_i$ [see (24) and (28) above], we note that

$$[\Phi - \Gamma] f = -\{f, h\} . \tag{41}$$

Thus (36) and (37) say that in a P transformation the terms $f_{k+\nu}$ (with $\nu < k$) are changed according to

$$\widetilde{f}_{k+\nu} = f_{k+\nu} - L_\nu (h_k) . \tag{42}$$

The operator $\mathcal{A} = L_0$, associated with the linear part $A = (Df)(0)$ of f ($f_1(x) = Ax$) is called the *homological operator*; it leaves the spaces $\mathcal{V}_k$ invariant, and it admits the natural decomposition

$$\mathcal{A} = \bigoplus_{m=0}^{\infty} \mathcal{A}^{(m)} , \tag{43}$$

where $\mathcal{A}^{(m)}$ is just the restriction of $\mathcal{A}$ to $\mathcal{V}_m$.

In the following, we will need to consider the adjoint $\mathcal{A}^+$ of the operator $\mathcal{A}$. For this we need to introduce a scalar product in the space $\mathcal{V}$. Actually, we can introduce a scalar product in each of the spaces $\mathcal{V}_k$ into which $\mathcal{V}$ decomposes.

A natural way of introducing such a scalar product [12] is to consider a basis $\{\xi_{\mu,s}\}$ in each of the finite-dimensional spaces $\mathcal{V}_k$, and then define $(\xi_{\mu,i}, \xi_{\nu,j}) = \delta_{\mu,\nu}\delta_{i,j}$. Such a basis function can be given in components by $\xi^i_{\mu,s} = [\prod_{i=1}^n x_i^{\mu_i}]\, \delta_{i,s}$, where μ are multiindices of modulus k, i.e. $\mu_i \in \mathbf{Z}_+$, $\sum_i \mu_i = k$. This will be referred to as the natural scalar product in $\mathcal{V}$.

---

[5] In some cases, it is more convenient to consider the so called Poincaré-Lie transformations rather than the Poincaré ones considered here (see the Appendix to this chapter). We stress that, as will be remarked in the Appendix, (37) also holds in that case.

Although this is a natural scalar product, it turns out that it is in many instances more convenient to consider a modification of it. This was introduced in [121] – following [31] – as $(\xi_{\mu,i},\xi_{\nu,j}) = \sum_{i=1}^n < \xi_{\mu,i},\xi_{\nu,i} >$, where $< x^\mu, x^\nu > \equiv \partial^\mu x^\nu$. Here, $\partial^\mu = \partial_1^{\mu_1} ... \partial_n^{\mu_n}$. It is easy to see that, in the present context, this is equivalent to defining $(\xi_{\mu,i},\xi_{\nu,j}) = \delta_{\mu,\nu}\delta_{i,j}(\mu!)$, where for the multiindex μ we have defined $\mu! = \prod_{i=1}^n (\mu_i!)$. Thus, this scalar product amounts to a modification, which can be interpreted as giving different weights to different orthogonal directions, of the previously defined one.

In the following, this will be referred to as the standard scalar product in $\mathcal{V}$. When considering adjoint operators these will have to be understood in terms of this scalar product.

What we need is the following classical result[6]; this can be proven by computation over the basis $\xi_{\mu,i}$, or see [190] for a simple proof.

Lemma 5. *If $\mathcal{A}$ is the homological operator associated with the matrix A, $\mathcal{A} = \{Ax, .\}$, then its adjoint $\mathcal{A}^+$ is the homological operator associated with the adjoint matrix A^+; i.e., $\mathcal{A}^+ = \{A^+x, .\}$.*

We now want to consider the (Poincaré) problem of classifying analytic vector fields in $\mathcal{M}$, or equivalently analytic vector functions $f \in \mathcal{V}$, up to equivalence under formal changes of coordinates, or under P transformations. In doing this, it is convenient to actually consider formal power series, and to discuss their convergence separately.

The advantage of considering P transformations is that, thanks to (35), we can proceed sequentially to normalize the terms $f_1, f_2, ...$, as at each stage the terms already normalized (those with $m < k$) are not changed any more.

If we denote by $\pi_k : \mathcal{V}_k \to \mathcal{V}_k$ the projection onto the range of $\mathcal{A}^{(k)}$, the equation for h_k

$$\mathcal{A}^{(k)}(h_k) = \pi_k f_k \tag{44}$$

always admits a solution in $\mathcal{V}_k$. Equation (44) is also called the *homological equation.*

The $h_k \in \mathcal{V}_k$ solutions to (44) will be of the form $h_k = h_k^* + \ell_k$, where

$$h_k^* = \mathcal{A}^+(\pi_k f_k) \in \text{Ran}\left[\left(\mathcal{A}^{(k)}\right)^+\right] \tag{45}$$

is unique, and ℓ_k is any function in $\text{Ker}\left[\mathcal{A}^{(k)}\right]$.

It should be stressed that, while adding a nonzero ℓ_k to h_k^* does not change the resulting $\widetilde{f}_k$, it could in general affect the terms of higher order.

We can then normalize X_f in the following recursive way, starting at $k = 1$: *i*) We write the homological equation (44) of order k; *ii*) we determine the

[6] We stress that this result only holds when the adjoint is taken with respect to the standard scalar product introduced above, and not, e.g., with respect to the "natural" scalar product we first introduced. This can be checked by considering the case A is a (2×2) matrix with zero entries except for $A_{12} = 1$.

unique h_k^* among its solutions (this amounts to a problem in linear algebra once we introduce the basis $\{e_i\}$ in $\mathcal{V}_k$); *iii)* we perform a P transformation with generator h_k^*, thus obtaining a new dynamical system, in normal form up to terms f_k. We can then repeat the procedure at successive orders, $k+1, k+2, ...$ up to any desired order N.

In this way, we are reduced to considering only systems characterized by the fact that

$$f_k \in \left[\text{Ran} \left(\mathcal{A}^{(k)} \right) \right]^{\perp} = \text{Ker} \left[\left(\mathcal{A}^{(k)} \right)^{+} \right] . \tag{46}$$

Such terms are also called *resonant.*

If the system $\dot{x} = f(x)$ has only resonant nonlinear terms, we say that it is in *Poincaré-Dulac normal form* [12,16,69,82]. If all the nonlinear terms of order up to q are resonant, we say that the system is in normal form up to order q. The discussion above shows that we have

Theorem 9 (Poincaré-Dulac). *Any analytic vector field X_f with $f(0) = 0$ can be formally taken to normal form by a formal series of Poincaré transformations.*

The name "resonant" is due to the relation existing between eigenvectors of $\mathcal{A}$ and resonance relations among eigenvalues of the matrix $A = (Df)(0)$ describing the linear part $f_0(x) = Ax$ of the system.

Let $\lambda_1, ..., \lambda_n$ be the eigenvalues of A, and let the corresponding eigenvectors[7] be $\xi^1, ..., \xi^n$. We have that the eigenfunctions of $\mathcal{A}^k$ are the monomials $\varphi_{(\mu,s)} = p(x)\xi_s$, where $p(x) = x^\mu$, with $\mu = (m_1, ..., m_n)$ a multiindex. Indeed, $Ap(x)\xi_s = p(x)\lambda_s\xi_s$; and $(A_{ij}x_j)(\partial p(x)/\partial x^i)\xi_s = (m \cdot \lambda)p(x)\xi_s$, where we have introduced the notation $(m \cdot \lambda) = \sum_i (m_i\lambda_i)$.

Thus, in conclusion,

$$\mathcal{A}\varphi_{(\mu,s)} = [(m \cdot \lambda) - \lambda_s] \ \varphi_{(\mu,s)} \ , \tag{47}$$

and the kernel of $L_0^{(k)}$ corresponds to monomials satisfying the *resonance condition*

$$\lambda_s \ = \ (m \cdot \lambda) \quad ; \quad \sum_i m_i = k \ . \tag{48}$$

We also say that (48) is a resonance condition of order $|m| = \sum_i m_i$.

Let us now consider the solution of the homological equation. By using the same basis of monomials $\varphi_{(\mu,s)} = x^\mu \xi_s$, we write for $g = \pi_k f_k$ and for the unknown $h = h_k$,

$$g = \sum v^{(\mu,s)} \varphi_{(\mu,s)} \quad , \quad h = \sum w^{(\mu,s)} \varphi_{(\mu,s)} \ . \tag{49}$$

The homological equation then gives immediately

$$w^{(\mu,s)} \ = \ \frac{v^{(\mu,s)}}{[(m \cdot \lambda) - \lambda_s]} \ . \tag{50}$$

---

[7] We assume here that A is semisimple, so that we have n eigenvectors.

Note that if we do not have an exact resonance, but $(m \cdot \lambda) - \lambda_s \simeq 0$, we have a *small denominator*, and correspondingly a very large coefficient in h_k. Such small denominators are responsible for divergencies in the normalizing series. In this book, we will mostly consider formal series, except in Chap. VI, where the problem of convergence of the normalizing transformation will be considered in some detail. We refer readers for this problem also to, e.g., [12,17,66,67,69,299].

Exercise 16. Consider the vector field $X = -x^2\partial_x - y\partial_y$, studied in Exercises 8,12 and 15. Is it in normal form ?

Exercise 17. Consider the linear vector field $X_A = -y\partial_x + x\partial_y$. Determine the most general (analytic) vector field X_f on $\mathbf{R}^2$ such that $(Df)(0) = A$, and X_f is in normal form.

Exercise 18. Same as Exercise 17, but first with $X_A = x\partial_x + y\partial_y$, and then with $X_A = x\partial_x + ny\partial_y$, with $n \in \mathbf{Z}$.

Exercise 19. Consider the vector field $X_f = (-y-x^2+x^3)\partial_x+(x+xy+xy^2)\partial_y$. Which of its terms are resonant, and which non-resonant ? Put it into normal form up to order three by the Poincaré procedure.

Exercise 20. Consider the special case of analytic vector fields on $\mathbf{R}^1$. Can any such vector field $X_f = f(x)\partial_x$ be taken into linear form by the Poincaré procedure ?

6. Birkhoff-Gustavson Normal Forms

In the case of a Hamiltonian system, we could consider the Hamiltonian vector field X_H corresponding to the Hamiltonian $H : \mathrm{T}_*M \to \mathbf{R}$, $X_H = \sum_i (\partial H/\partial p_i)(\partial/\partial q^i) - (\partial H/\partial q^i)(\partial/\partial p_i)$, and apply the Poincaré-Dulac theory, preferably using only the P transformations which preserve the Hamiltonian (symplectic) structure in T_*M.

However, as was observed by Birkhoff, it is more convenient to work directly with the Hamiltonians H. One can then construct a theory for the normal form of Hamiltonians that is essentially equivalent (and parallel) to the theory of normal forms for vector fields. Such a theory was developed by Birkhoff (who worked the non-resonant case, as Poincaré had done for vector fields) and Gustavson (who extended Birkhoff's theory to the resonant case, as Dulac had done for Poincaré theory) (see [13,17,174,246,247,317]).

The similarity among the two theories is not accidental, nor only formal. It amounts to an underlying common algebraic structure, naturally set in terms of *filtrations of Lie algebras* [61,62,65]. We will not discuss this relation here.

Let $M \subseteq \mathbf{R}^\ell$ be a smooth manifold, and let (q,p) be coordinates in T_*M. For notational convenience, we will also consider coordinates $x_i \in \mathrm{T}_*M \subseteq \mathbf{R}^{(2\ell)} \equiv \mathbf{R}^n$ given by $x_i = q_i$ for $1 \le i \le \ell$ and $x_i = p_i$ for $\ell < i \le 2\ell$.

Let us now consider an analytic Hamiltonian[8] of the form

[8] By "Hamiltonian" we will always mean an analytic one.

$$H(q,p) = H_0 \ + \sum_{k=1}^{\infty} H_k \tag{51}$$

where H_m is an homogeneous polynomial of order[9] $m+2$,

$$H_m(aq, ap) = a^{m+2} H_m(q,p) \ . \tag{52}$$

The (antisymmetric) Poisson bracket $\{.,.\} : \mathcal{F} \times \mathcal{F} \to \mathcal{F}$ is defined[10] as

$$\{F, G\} = \sum_{i=1}^{\ell} \left[\frac{\partial F}{\partial q_i} \frac{\partial G}{\partial p_i} - \frac{\partial F}{\partial p_i} \frac{\partial G}{\partial q_i} \right] \ . \tag{53}$$

Note that $\{.,.\} : \mathcal{F}_a \times \mathcal{F}_b \to \mathcal{F}_{a+b}$. In particular, this means that $\{H_0, .\} : \mathcal{F}_k \to \mathcal{F}_k$.

To any $H \in \mathcal{F}$ we associate the vector field X_H on $T_* M$ generated by H according to the Hamilton equations[11], i.e.

$$X_H = \frac{\partial H}{\partial p_i} \frac{\partial}{\partial q_i} - \frac{\partial H}{\partial q_i} \frac{\partial}{\partial p_i} \equiv \dot{q}_i \frac{\partial}{\partial q_i} + \dot{p}_i \frac{\partial}{\partial p_i} \tag{54}$$

Thus, the time evolution of any $F \in M$ along the dynamics generated by H is given simply by $X_H(F)$, and we have

$$X_H(F) = \{F, H\} = \dot{q}_i \frac{\partial F}{\partial q_i} + \dot{p}_i \frac{\partial F}{\partial p_i} \ = \ \left(\frac{\mathrm{d}F}{\mathrm{d}t} \right)_H \tag{55}$$

and, as is well known, for any $F, G \in \mathcal{F}$ we have

$$X_G(F) = \{F, G\} = -\{G, F\} = -X_F(G) \ . \tag{56}$$

It should be stressed that the vanishing of $\{F, G\}$ and the vanishing of $[X_F, X_G]$ are *not* equivalent. Indeed, by direct computation we have that

$$[X_F, X_G] = \left[-\frac{\partial \{F, G\}}{\partial p_j} \right] \frac{\partial}{\partial q_j} + \left[\frac{\partial \{F, G\}}{\partial q_j} \right] \frac{\partial}{\partial p_j} \tag{57}$$

so that $[X_F, X_G] = 0$ only implies $\{F, G\} = c$. By the same computation, we also have that

---

[9] This apparently unnatural choice of notation comes from the desire of denoting by H_0 the quadratic term, representing the integrable system of which the full H is seen as a perturbation.

[10] Obviously, this should not be confused with the bracket $\{.,.\}$ on $\mathcal{V}$ given by (38), and denoted below as $\{.,.\}_\mathcal{V}$.

[11] With the notation employed above for the general case, this would be written as $X_{J\nabla H}$, where J is the standard 2ℓ dimensional symplectic matrix: indeed (54) corresponds to the evolution given by $\dot{x} = J\nabla H$; adopting this special notation for the Hamiltonian case should not induce any confusion. Notice also that for $f = J\nabla F$ and $g = J\nabla G$ we have $\{f, g\}_\mathcal{V} = -J\nabla\{F, G\}$.

$$\{F,G\} = K \iff [X_F, X_G] = -X_K \ . \tag{58}$$

Let us now come back to the Hamiltonian (51), and let us consider a change of coordinates which corresponds to the action of the Lie operator[12] [47] associated to a homogeneous function $F \in \mathcal{F}_m$. This Lie operator, denoted by $\mathcal{L}_F : \mathcal{F} \to \mathcal{F}$, is defined as the time-one action of the flow under $(-X_F)$, i.e. with

$$\mathrm{ad}_F(.) = \{F, .\} \tag{59}$$

we define

$$\mathcal{L}_F(.) = \exp[\mathrm{ad}_F](.) = \sum_{r=0}^{\infty} \frac{1}{r!} \left(\mathrm{ad}_F\right)^r (.) \tag{60}$$

It should be stressed that, when performing computations only up to order m – as we will do in the following – the action of $\mathcal{L}_F$ reduces to the change of coordinates

$$q = Q + \partial F/\partial P \ ; \ \ p = P - \partial F/\partial Q \ . \tag{61}$$

The advantage of considering Lie operators lies in that, being defined in terms of Poisson brackets, these generate a transformation that is guaranteed to be canonical at all orders; moreover, these simplify some step in the Birkhoff normalization procedure, as the inversion of certain operators is now immediately guaranteed [47]. We will not dwell on these topics.

By direct computation, one has that under the change of coordinates generated by $\mathcal{L}_F$ – or, equivalently, under (61) – the Hamiltonian H is changed to $\widetilde{H} = \mathcal{L}_F(H)$ given, up to terms in M_{m+1} (unspecified terms in $\mathcal{F}_k$ are denoted by $\mathcal{R}_k$ below), by

$$\begin{aligned} \mathcal{L}_F(H) \equiv \widetilde{H} =& H + \{F_m, H\} + \mathcal{R}_{2m} = \\ &= H_0 + \sum_{k=1}^{\infty} H_k + \{F_m, H_0\} + \sum_{k=1}^{\infty} \{F_m, H_k\} + \mathcal{R}_{2m} = \\ &= H_0 + \sum_{k=1}^{m} H_k \ + \{F_m, H_0\} + \mathcal{R}_{m+1} \ . \end{aligned} \tag{62}$$

This means that with the change of coordinates generated by F_m, which we call a Birkhoff (B) transformation, the Hamiltonian (51) is transformed to a new one

$$\widetilde{H} = H_0 + \sum_{k=1}^{\infty} \widetilde{H}_k \tag{63}$$

where in particular $\widetilde{H}_k = H_k$ for $k < m$; and

$$\widetilde{H}_m = H_m + \{F_m, H_0\} \ . \tag{64}$$

We introduce the *homological operator* ζ associated with H_0, defined as

[12] This is analogous to the Lie transforms discussed in the Appendix.

$$\zeta(.) = \{H_0, .\} \tag{65}$$

(note that $\zeta : \mathcal{F}_k \to \mathcal{F}_k$), so that (64) also reads

$$\widetilde{H}_m = H_m - \zeta(F_m) \ . \tag{66}$$

It is clear that by a series of B transformations, H can be brought into

$$\widehat{H} = H_0 + \sum_{k=1}^{\infty} \widehat{H}_k \tag{67}$$

where all the $\widehat{H}_k$ satisfy

$$\widehat{H}_k \in [\mathrm{Ran}(\zeta)]^{\perp} = \mathrm{Ker}\left(\zeta^+\right) \tag{68}$$

with the adjoint to be meant with respect to a scalar product that is analogous to the standard scalar product considered in the Poincaré case; this is briefly discussed below. We will denote by π_ζ the projection $\pi_\zeta : \mathcal{F} \to \mathrm{Ran}(\zeta)$.

Here we assume that $[\zeta, \zeta^+] = 0$, for the sake of simplicity [in this way $\mathrm{Ker}(\zeta) = \mathrm{Ker}(\zeta^+)$]. In Chap. V we will have to consider the general case as well.

Notice that since $\zeta : \mathcal{F}_k \to \mathcal{F}_k$, we can naturally write $\zeta = \oplus_{k=0}^{\infty} \zeta_k$, with ζ_k the restriction of ζ to $\mathcal{F}_k$. All the $\mathcal{F}_k$ are of finite dimension, which simplifies the analysis of ζ, as this can be divided into the analysis of the ζ_k. Thus, the above analysis can be made completely rigorous, along the lines of the discussion for the Poincaré theory.

It is important to emphasize that the reduction from H to its normal form $\widehat{H}$ (67) requires one to use, at each step, the appropriate F_m. This is obtained as the solution of the *homological equation*

$$\zeta(F_m) = \pi_\zeta H_m \ . \tag{69}$$

It is also relevant, although obvious, that the solutions to these are determined up to terms

$$\delta F_m \in \mathrm{Ker}(\zeta) \ . \tag{70}$$

Note that, rather than defining the Lie transform operator $\mathcal{L}_{F_m}$, we could have simply considered the change of coordinates (61) leading to the same results.

A system in the form (67) which satisfies (68) is said to be in *Birkhoff-Gustavson normal form*. We have immediately that

Lemma 6. *The Hamiltonian* $H = \sum_{m=0}^{\infty} H_m$ *is in Birkhoff-Gustavson normal form if and only if* $\{H_0, H_m\} = 0$.

The discussion conducted above shows that in the Hamiltonian case we have the following analogue of the Poincaré-Dulac theorem:

Theorem 10 (Birkhoff). *Any analytic Hamiltonian* H *with* $H(0,0) = (DH)(0,0) = 0$ *can be formally taken to normal form by a formal series of Birkhoff transformations.*

This means, in particular, that the normal form expansion of any Hamiltonian with quadratic part H_0 can be obtained by studying, order by order, the solution to $\zeta(\widehat{H}_m) \equiv \{H_0, \widehat{H}_m\} = 0$ or, equivalently, the (Hamiltonian) resonance conditions [12,13] for H_0.

As in the Poincaré-Dulac theory, a crucial role is played by the adjoint of the homological operator, ζ^+. We stated above that this adjoint must be meant with respect to a standard scalar product in $\mathcal{F}$. A natural basis in $\mathcal{F}$ is provided by the monomials, which we write in multiindex notation as $x^\mu = x_1^{\mu_1}...x_n^{\mu_n} \in \mathcal{F}_{|\mu|}$. The scalar product is then defined on these, in analogy with the scalar product defined above in $\mathcal{V}$, as $(x^\mu, x^\nu) = \delta_{\mu,\nu}(\mu!)$, where again $\mu! = \prod_{i=1}^n (\mu_i!)$. We refer to this as the standard scalar product in $\mathcal{F}$, and adjoints of operators will be understood with respect to this scalar product..

In particular, it is relevant to know whether ζ is a normal operator, i.e. if $[\zeta, \zeta^+] = 0$. The answer to this is provided by the following classical result (see Chap. V for a proof):

Lemma 7. *Let $H_0 = (1/2)C_{ij}x_ix_j$, $x \in \mathrm{T}_*M$, be a quadratic Hamiltonian, in which we can take $C_{ij} = C_{ji}$. Let $\zeta = \{H_0, .\}$ be the associated homological operator. Let J be the standard symplectic matrix in T_*M. Then, ζ is normal if and only if $A = JC$ is a normal matrix.*

Exercise 21. Consider the Hamiltonian (on $\mathrm{T}_*\mathbf{R}^2$) $H = (1/2)(x^2 + y^2 + u^2 + v^2) + \left[(1/2)(x^2 + u^2) + (xy + uv)^2\right]^2$. Is it in normal form ?

Exercise 22. Consider the quadratic Hamiltonian $H_0 = (1/2)(p_1^2 + q_1^2) + (1/2)(p_2^2 + \omega^2 q_2^2)$. Determine the most general Hamiltonian H on $\mathrm{T}_*\mathbf{R}^2$ which is in normal form and has a quadratic part H_0, depending on the value of $\omega \in \mathbf{R}$.

Exercise 23. The same as in Exercise 22, but with $H_0 = (1/2)(p_1^2 + q_1^2) + (1/2)(p_2^2 + nq_2^2)$ with $n \in \mathbf{Z}$.

Exercise 24. The same as in Exercise 22, but with $H_0 = (1/2)(p_1^2 + p_2^2)$ (free particle).

Exercise 25. When we consider action-angle variables $(I_1, ..., I_n; \varphi_1, ..., \varphi_n)$, the quadratic Hamiltonian $H_0 = (1/2)\sum_{i=1}^n (p_i^2 + \omega_i q_i^2)$ is written as $H_0 = \sum_{i=1}^n I_i$, so that the corresponding equations of motion are $\dot{I}_i = 0$, $\dot{\varphi}_i = \partial H / \partial I_i$. Let us consider the general Hamiltonian $H(q,p)$ having H_0 as quadratic part. What can we say about the normal form expansion, respectively, in the action-angle and in the (q,p) coordinates ? Does this normal form expansion permit us to identify functions which, for general H, vary slowly in time in some subset of the phase space ?

Exercise 26. Consider the Hamiltonian $H = (1/2)(x^2 + u^2) + (xy + uv)^2$. Transform it into normal form up to order three.

Exercise 27. Consider the special case $M = \mathbf{R}^1$, so that $q \in \mathbf{R}^1$, $p \in \mathbf{R}^1$. Can any Hamiltonian $H(q,p)$ be reduced to its quadratic part by the Birkhoff normalization procedure ?

7. Bifurcation Theory

In this volume we will not consider specifically bifurcation theory [3,16,81,104, 173,191,231,279,287]. On the other hand, this is now a widely used tool in perturbative analysis of nonlinear systems and, moreover, it is probably the first field in which the interplay between symmetry and perturbative methods was explicitly – and productively – considered by Ruelle, Sattinger, and then many others [80,105,125-131,143,169,190,236,278,288,289]. We will thus, for completeness, give a short introduction to this theory.

Roughly speaking, bifurcation theory considers the situation in which, as a control parameter is varied, a solution u_0 loses stability and new stable solutions u_n appear, initially very near to u_0. In physical terms, this corresponds to a *second order phase transition.* Let us now be more precise.

We consider an evolution PDE in some function space, which we assume for ease of discussion to be a Hilbert space H with some given scalar product and metric. We write the evolution PDE as

$$\frac{\partial u(x,t)}{\partial t} = G[\mu, u(x,t)] \ , \tag{71}$$

where μ is an external real control parameter, $u : \mathbf{R}^n \times \mathbf{R} \to \mathbf{R}^m$ and G will depend on u and spatial (x) derivatives of u. We assume G to be smooth in both μ and u. Equation (71) has to be supplemented with boundary conditions, specifying the function space H to which u is required to belong.

We suppose, moreover, that $u(x,t) = 0$ (also denoted as u_0) is a solution of (71), i.e. that $G[\mu, u_0] = 0$ for any μ (by rescaling u, any other given function which is a solution for all μ would be equivalent).

We consider then the linearization $L(\mu)$ of G at u_0, $L(\mu) = (DG)[\mu, u_0]$. This is assumed to be a Fredholm operator of index zero. We have to make some further assumptions on the spectrum of $L(\mu)$, which will guarantee that we have indeed a bifurcation.

We assume that for $\mu < \mu_0$ (we can assume $\mu_0 = 0$ with no loss of generality) the whole spectrum lies in the left half of complex plane, so that for $\mu < \mu_0$ the solution u_0 is stable. This same condition also implies, due to the implicit function theorem, that for $\mu < \mu_0$ the solution u_0 is isolated, i.e. there is some $\rho(\mu) > 0$ such that if u_1 is another solution, then $||u_0 - u_1|| \geq \rho(\mu)$.

We also assume that for $\mu = \mu_0$ a finite set of eigenvalues of $L(\mu)$ crosses the imaginary axis with positive velocity, i.e. there are $\lambda_1(\mu), ...\lambda_k(\mu)$, eigenvalues of $L(\mu)$, such that

$$\mathrm{Re}[\lambda_i(\mu_0)] = 0 \quad , \qquad \frac{\mathrm{dRe}[\lambda_i(\mu)]}{\mathrm{d}\mu}\Big|_{\mu_0} > 0 \ , \tag{72}$$

while all the remaining part of the spectrum is at a finite distance from the imaginary axis (gap condition). We denote by $E(\mu)$ the linear space spanned by the eigenvectors corresponding to these critical eigenvalues.

This means that u_0 loses stability in a finite number of directions.[13] Notice that at μ_0 the implicit function theorem does not forbid any more that there are solutions u_1 as near as we wish to u_0.

If there exists some family of functions $u^{(\mu)}(x,t)$, defined for $\mu \in \mathcal{M}$, where $\mathcal{M}$ is a neighborhood of μ_0 in $\mathbf{R}$ (possibly with $\mu \in \partial\mathcal{M}$, e.g. defined only for $\mu \geq \mu_0$) such that $\partial_t u^{(\mu)} = G[\mu, u^{(\mu)}]$ (that is, (71) has $u^{(\mu)}$ as a solution, $\forall \mu \in \mathcal{M}$), and such that

$$\lim_{\mu \to \mu_0} u^{(\mu)}(x,t) = u_0(x,t) , \tag{73}$$

we say that the family of solutions $u^{(\mu)}$ *bifurcates* from u_0 at the *bifurcation point* $\mu = \mu_0$.

A generalization of the center manifold theorem guarantees then that there is an invariant (center-unstable) k-dimensional local manifold $W_0(\mu)$, smoothly depending on μ, tangent in u_0 to $E(\mu)$. Moreover, this will be transversally hyperbolic (and attractive) for μ in a neighborhood of μ_0, due to the gap condition and smoothness of $L(\mu)$. Thus, the asymptotic behavior of (71) in a neighborhood of u_0 is described by the flow restricted to the finite dimensional manifold W_0.

Consider a neighborhood $U \subset H$ of W_0, and on this take local coordinates $u = (v, w)$, such that W_0 corresponds to points $(0, w)$. The flow on W_0 will be described by a finite dimensional evolution equation

$$\frac{\mathrm{d}w^i}{\mathrm{d}t} = f^i(w) , \tag{74}$$

called the *bifurcation equation*. We then have

Theorem 11 (Bifurcation theorem). *There exist sufficiently small neighborhoods $\mathcal{M} \subset \mathbf{R}$ of μ_0 and $U \subset H$ of u_0, $U_W = W_0 \cap U \subset W_0$, such that for $t \to \infty$ the solutions of the original problem lying in U and the solutions of the bifurcation equation lying in U_W are in one to one correspondence.*

For clarification, it can be useful to consider the simple case of stationary solutions, in which one can apply the Lyapounov-Schmidt construction for the bifurcation equation.

Let $P : H \to E(\mu)$ be the orthogonal projection on $E(\mu)$, and let $Q = (I - P)$. Apply both P and Q on (71) and look for stationary solutions, i.e. require that $\dot{u} = \dot{v} = \dot{w} = 0$. This gives two equations,

$$\begin{aligned} QG[\mu, v + w] &= 0 \\ PG[\mu, v + w] &= 0 . \end{aligned} \tag{75}$$

Due to the implicit function theorem, we are guaranteed that the first one can be locally solved for v as a function of w; call the solution $v = h(\mu, w)$. We can then substitute in the second equation, and get the bifurcation equation as

[13] In physical terms, there are k critical modes.

$$f(\mu, w) \equiv PG[\mu, w + h(\mu, w)] = 0 \ . \tag{76}$$

Exercise 28. Consider the following ODE in $\mathbf{R}^1$: $\dot{x} = \mu x - x^3$. Check that for this there is a bifurcation of stationary solutions at $\mu_0 = 0$. Check also that there are two branches $x_\pm(\mu)$ of bifurcating stationary solutions, defined for $\mu \geq \mu_0$, and that these are stable for all $\mu \geq \mu_0$. Is there any relation between the fact that the ODE is invariant for $x \to -x$ and the fact that $x_\pm(\mu) = -x_\mp(\mu)$?

Exercise 29. Consider the equations

$$\begin{aligned} \dot{x} &= \mu x - y - (x^2 + y^2)x \\ \dot{y} &= x + \mu y - (x^2 + y^2)y \ . \end{aligned}$$

Show (possibly using a rotating frame of reference and the Lyapounov-Schmidt method) that there are periodic solutions $\xi_\vartheta^{(\mu)}(t) \in \mathbf{R}^2$, given by rotation with constant angular velocity $\omega = 1$ on circles of radius $r(\mu) = \sqrt{\mu}$ bifurcating from the null solution at $\mu_0 = 0$. Show that the branches of bifurcating solutions are indexed by a phase $\vartheta \in S^1$, and that these are transformed into each other by $\xi_{\vartheta+\tau}^{(\mu)}(t) = \xi_\vartheta^{(\mu)}(t + \tau)$. Discuss the relation between the rotational invariance of the original equation and this property of the bifurcating solutions, paying care to avoid confusion between rotations in space and time shifts along the periodic solution.

Exercise 30. Consider the case when the original problem is a map from $\mathbf{R} \times \mathbf{R}^n$ into $\mathbf{R}^n$, i.e. we have $G(\mu, u) = 0$ with $u \in \mathbf{R}^n$, $G(\mu, u) \in \mathbf{R}^n$. Show that if $G(\mu, u)$ satisfies $G^i(\mu, Tu) = T^i_{\ j} G^j(\mu, u)$, with T a matrix, then necessarily $E = \text{Ker}\,(L_{\mu_0})$ is invariant under T. Let S be the restriction of T to E. Consider now the Lyapounov-Schmidt procedure shown above, and prove that the bifurcation equation $F(\mu, w) = 0$ satisfies $F^i(\mu, Sw) = S^i_{\ j} F^j(\mu, w)$.

Appendix. Lie Transforms

In the study of Poincaré normal forms, we have considered transformations of the kind

$$y = \Phi(x) \equiv x + h(x) \tag{A.1}$$

where $\Phi : M \to M$ is a near-identity diffeomorphism of M; this means in particular that it can be expressed as the time-one map for the flow under some vector field $H \in \mathcal{M}$. In a number of cases, it would be more convenient to deal directly with such vector fields rather than with the finite transformation Φ. We are going to briefly discuss this approach, and its relation with the one discussed above.

Let the vector field[14] $H \equiv X_h \in \mathcal{M}$ be given, in the x coordinates, by

$$H = h^i(x) \frac{\partial}{\partial x^i} \ . \tag{A.2}$$

We denote by Ψ the local[15] flow under H, i.e.

$$\frac{\mathrm{d}}{\mathrm{d}s} \Psi(s;x) \ = \ H\left(\Psi(s;x)\right) \ . \tag{A.3}$$

We also use exponential notation, i.e. we write

$$\Psi(s;x) \ = \ e^{sH} x \ . \tag{A.4}$$

With this, we decide to denote $\Psi(1;y)$ as x, i.e. our change of coordinates will be defined to be

$$x = \Psi(1;y) = \left[e^{sH} y \right]_{s=1} \ , \tag{A.5}$$

and conversely we have

$$y = \Psi(-1;x) = \left[e^{-sH} x \right]_{s=1} \ . \tag{A.6}$$

Now, let us consider another vector field X on M, describing the dynamical system we are interested in. Namely, if we study

$$\dot{y}^i = f^i(y) \ , \tag{A.7}$$

we consider the vector field given, in the y coordinates, by

$$X = f^i(y) \ \frac{\partial}{\partial y^i} \ . \tag{A.8}$$

[14] We use here H rather than X_h only to simplify notation and to avoid to have too many indices, as it will repeatedly appear as the argument of an exponential. This H should not, obviously, be confused with a Hamiltonian function.

[15] If $h(x)$ has some zero, Ψ will only be a semiflow, and not a flow, globally in time; however, for finite times H defines a local flow.

This does also generate a (local) flow, i.e. for any $y_0 \in M$ we have a one-parameter family $y(t) \equiv \Gamma(t; y_0) \in M$ such that

$$\frac{\mathrm{d}y(t)}{\mathrm{d}t} = X(y) \,. \tag{A.9}$$

By (A.5), this also defined a one-parameter family $x(t) \in M$, which will satisfy

$$\frac{\mathrm{d}x(t)}{\mathrm{d}t} = \widetilde{X}(x) \tag{A.10}$$

for some vector field $\widetilde{X} \in \mathcal{M}$. We have to determine $\widetilde{X} = \widetilde{f}^i(x)\partial/\partial x^i$.

We write, using the exponential notation, $y(t+\delta t) = e^{\delta t X} y(t)$. Therefore,

$$x(t+\delta t) = \left[e^{sH} \ e^{(\delta t)X} \ y(t) \right]_{s=1} \,. \tag{A.11}$$

Using $y(t) = e^{-H}x(t)$, we have

$$\frac{x(t+\delta t) - x(t)}{\delta t} = \left[e^{sH} \ \left(e^{(\delta t)X} - I \right) \ e^{-sH} \ x(t) \right]_{(s=1)} \,, \tag{A.12}$$

and therefore

$$\widetilde{X} = \left[e^{sH} \ X \ e^{-sH} \right]_{(s=1)} \,. \tag{A.13}$$

We call this transformation the Poincaré-Lie transformation generated by h. Notice that this yields, up to order one in s, and therefore if $h = h_k$, up to order terms in V_k, just the same transformation as the Poincaré transformation with the same generator h.

The $\widetilde{X}$ can be given (for arbitrary s), in terms of the Baker-Campbell-Haussdorf formula, as

$$\widetilde{X} = \sum_{k=0}^{\infty} \frac{(-1)^k s^k}{k!} \ X^{(k)} \,, \tag{A.14}$$

where $X^{(0)} = X$, and

$$X^{(k+1)} = \left[X^{(k)}, H \right] \,. \tag{A.15}$$

As this, or equivalently (A.13), define a transform for any s, we can consider the continuous family of vector fields $X_{(s)}$ given by

$$X_{(s)} = f^i_{(s)}(x_{(s)}) \frac{\partial}{\partial x^i_{(s)}} = e^{sH} X e^{-sH} \,, \tag{A.16}$$

where obviously $X_{(0)} = X$, $X_{(1)} = \widetilde{X}$. The $f_{(s)}$ will also change continuously with s, according to

$$\frac{\mathrm{d}f^i_{(s)}}{\mathrm{d}s} = \{h, f_{(s)}\}^i \,. \tag{A.17}$$

If now $h = h_k \in V_k$, and we expand f in homogeneous terms, $f = \sum f_m$ (writing $f(s) \equiv f_{(s)}$ to avoid confusion), the above formula yields immediately

$$\frac{\mathrm{d}f_m(s)}{\mathrm{d}s} = \begin{cases} \{h_k, f_{m-k}\} & \text{for } m \geq k \\ 0 & \text{for } m < k \ . \end{cases} \tag{A.18}$$

For $m = k$, we just have the action of the homological operator.

In the same vein and with the same notation, we can consider the action of the vector field X_A corresponding to the linear part Ax of the vector field X. With the same computation, we have that now

$$\widehat{X}_{(\lambda)} = \widehat{f}^i(x)\frac{\partial}{\partial x^i} = e^{\lambda X_A} \ X \ e^{-\lambda X_A} \tag{A.19}$$

satisfies $\mathrm{d}\widehat{X}_{(\lambda)}/\mathrm{d}\lambda = [X_A, X]$, i.e.

$$\frac{\mathrm{d}\widehat{f}(\lambda)}{\mathrm{d}\lambda} = \{Ax, \widehat{f}\} \ , \tag{A.20}$$

and therefore we can also write the action of the homological operator $\mathcal{A}$ associated with A as

$$\mathcal{A}(X_f) = \left[\frac{\mathrm{d}}{\mathrm{d}\lambda}e^{\lambda X_A} \ X_f \ e^{-\lambda X_A}\right]_{(\lambda=0)} \ . \tag{A.21}$$

This form is quite useful in simplifying a number of proofs.

Similar considerations would also apply for what concerns Birkhoff normal forms, and Birkhoff transformations.

III. Symmetries of Dynamical Systems

The application of symmetry methods for differential equations for dynamical systems (as defined in Chap. II) presents some special aspects [146,305], due to the structure of the differential equations under study. In a sense, to be clear in the following, when having to deal with the specially simple class of equations consisting of first order autonomous ODEs, one is faced with a specially difficult problem when trying to determine its symmetries. Indeed one has, in general, an infinite dimensional symmetry algebra. Moreover, as we will see below, the structure of this is related to the constants of motion for the system.

Thus, one studies the symmetry of the system to learn more about its behaviour, periodic and otherwise noteworthy solutions, conserved quantities, etc., but on the other hand information about these solutions, conserved quantities, etc. is needed to unravel the structure of the symmetry algebra!

The two problems are strongly related, and indeed we will see that determining the full symmetry algebra of the dynamical system is not any easier than determining its most general solution. Luckily, on the one hand we can make good use of symmetry methods based on a partial knowledge of the symmetry algebra, and on the other hand in the study of dynamical systems we are often satisfied – or at least we are used to be satisfied, as we cannot usually go beyond this stage – with qualitative and/or semiquantitative, and anyway partial, descriptions of its behaviour.

Another important fact is that we can proceed in a perturbative way, not only in the study of the dynamical system (Chap. II), but also in the determination of its symmetries. Moreover, an approximate determination of a symmetry – or even the determination of an approximate symmetry[1] (see below) – gives useful information on the perturbative expansion of the dynamical systems. Actually, as we will argue below, once we are studying the dynamical system by a perturbative approach, approximate systems are at least as useful as exact ones [94,159].

The specific structure of dynamical systems will lead to the above mentioned relation between symmetries and constants of motion, and the infinite dimensional structure of the symmetry algebra (actually, as we will see below, it would be more appropriate to speak of the *symmetry module*), but it does also naturally lead to consider a more restricted class of symmetries, i.e. those which leave

[1] This includes not only the case of a vector field that is obtained as an approximation of an exact symmetry, but also that of a vector field such that its commutator with the dynamical vector field is in some sense "small", but which is not "near" an exact symmetry.

globally invariant the class of dynamical systems ODEs. From a physical point of view, these are the symmetries which "recognize" the special role of time, i.e. of the independent variable. Considering these will also focus our attention on the trajectories of the dynamical system (rather than on full solutions, i.e. integral curves in space-time), and this will in turn lead to relations between such symmetries and the topology of trajectories for solutions to the dynamical system.

In the final section, we will consider a topic that lies somewhat out of the main line of development of these notes, but is, however, (in our opinion) very relevant, i.e. *approximate symmetries.* For these, we will only introduce the basic concepts and study their algebraic structure, which is equivalent to that of exact symmetries. Applications of these will not be considered, but are discussed in [94].

In the Appendix, we discuss some points concerning the module structure of $\mathcal{G}_X$. This will be further discussed in the next chapter, but here we want to warn the reader about some relatively common misunderstandings, and clarify some points.

1. Symmetries of Dynamical Systems

We will write, as in Chap. II, a dynamical system (DS) in the form

$$\dot{x} \;=\; f(x) \quad ; \; x \in M \; ; \; f : M \to \mathrm{T}M \tag{1}$$

where $M \subseteq R^n$ is a smooth manifold (regularly embedded in R^n), x are cartesian coordinates in R^n, and f a smooth function. We also associate to (1) the vector field

$$X_f \;=\; f^i(x)\,\frac{\partial}{\partial x^i} \quad . \tag{2}$$

According to our general procedure, we look for symmetry vector fields[2] in the form

$$X \;=\; \tau(x,t)\,\partial_t \;+\; \varphi^i(x,t)\,\partial_i \;, \tag{3}$$

where we have written $\partial_i \equiv (\partial/\partial x^i)$. The first prolongation of this is

$$X^{(1)} \;=\; X + \Phi^{(i)}\frac{\partial}{\partial \dot{x}^i} \quad , \tag{4}$$

and according to the general prolongation formula the coefficients $\Phi^{(i)}$ are given by

$$\Phi^{(i)} \;=\; \left[\partial_t\varphi^i + \dot{x}^j\partial_j\varphi^i - \dot{x}^i(\partial_t\tau + \dot{x}^j\partial_j\tau)\right] \; . \tag{5}$$

Thus, by applying $X^{(1)}$ on the ODE (1) describing the DS, and by requiring the vanishing of this on the solution to (1), we get a rather complicate system of linear PDEs for τ and the φ's, which can be written explicitly as

[2] Note we are restricting, as nearly always in this volume, our attention to continuous symmetry; for dicrete symmetries see e.g. [202].

$$\frac{\partial \varphi^i}{\partial_t} + f^j(x)\frac{\partial \varphi^i}{\partial x^j} - \frac{\partial f^i(x)}{\partial x^j}\varphi^j = f^i(x)\left(\frac{\partial \tau}{\partial t} + f^j(x)\frac{\partial \tau}{\partial x^j}\right) . \tag{6}$$

This is in general, clearly, an underdetermined system: indeed, we have n equations for the $(n+1)$ functions $\varphi^1, ..., \varphi^n; \tau$.

2. Lie-Point Time-Independent Symmetries

When we apply a symmetry transformation like (3) – or more precisely, its first prolongation (4) – on a DS (1), we do not, in general, again obtain a DS.

This is a general situation, i.e. we have a similar problem also for general ODEs, and for PDEs, when we allow dependent and independent variables to mix through the symmetry transformation. In order to guarantee that an evolution equation (in this case an ODE) is transformed into an equation of the same class[3], we have to require that the time transformation does not depend on spatial independent variables (absent in this case), nor on the dependent variables. In the present notation, this means that we have to ask that τ be a function of t alone, i.e. that the effect of X on the time variable amounts to a reparametrization of time.

It should be emphasized that this requirement is sufficient to guarantee that a (system of) first order ODE is transformed into a system of the same kind, but *not* that an autonomous system remains autonomous. By direct inspection, we see that even with this hypothesis (i.e. assuming $\tau_j = 0$) the system is changed, up to first order terms, into

$$\dot{x}^i = f^i(x) + \varepsilon\left[\left(\varphi^j\partial_j f^i - f^j\partial_j\varphi^i\right) + \left(\tau_t f^i - \partial_t\varphi^i\right)\right] , \tag{7}$$

which in general is not an autonomous equation.

The condition for X to be a symmetry, i.e. to leave the DS unchanged, can be read directly from (7) to be, in this case,

$$\partial_t\varphi^i - \tau_t f^i = \left(\varphi^j\partial_j\right) f^i - \left(f^j\partial_j\right)\varphi^i . \tag{8}$$

The right hand side of this has a special relevance in all the theory of symmetries for dynamical systems. Thus we have a special notation for it, i.e. we define (Chaps. I and II) the Lie-Poisson bracket $\{.,.\} : \mathcal{M}\times\mathcal{M}\to\mathcal{M}$ (where $\mathcal{M}$ represents the algebra of, now possibly time-dependent, vector fields on M) by

$$\{f,g\} = (f\cdot\nabla)\,g - (g\cdot\nabla)\,f . \tag{9}$$

Note that (as already noted in Chap. I) this is just the commutator of vector fields read on their components. Indeed, if we have vector fields $X_f = f^i\partial_i$ and

[3] It should be stressed that we require this to be true for the full group of transformations generated by the symmetry vector field X, i.e. for all values of the group parameter λ in the transformation $e^{\lambda X}$.

$X_g = g^i \partial_i$, then their commutator $[X_f, X_g] = X_h = h^i \partial_i$ has components given by $h = \{f, g\}$.

In this notation, (8) reads simply

$$\varphi_t - \tau_t f = \{\varphi, f\} . \tag{10}$$

If we disregard the possibility of time reparametrizations, i.e. we require that our symmetry transformations do not affect the temporal coordinate, we have $\tau \equiv 0$, and (10) reduces to

$$\varphi_t = \{\varphi, f\} . \tag{11}$$

If we allow a nonzero τ (with $\tau_j = 0$ for all j), then we can define $\psi(x,t) = \varphi(x,t) - \tau(t) f(x)$, and (10) is then rewritten as

$$\psi_t = \{\psi, f\} . \tag{12}$$

Thus, once we impose the condition $\tau = \tau(t)$, we can limit ourselves to considering symmetries with $\tau = 0$. From now on we will indeed proceed in this way.

The reader familiar with the language of symmetry methods for differential equations will recognize at once that we are in this way essentially considering the *evolutionary representative* [262] of the vector field X.[4]

However, a vector field of this form, i.e. of the form $X = \varphi^i(x,t)\partial_i$, is not in general transforming dynamical systems into dynamical systems, although it can of course leave unchanged a given class of dynamical systems. To be sure that the class of dynamical systems is left globally invariant, we should also ask that $\varphi_t = 0$ (or $\psi_t = 0$ if we are dealing with the evolutionary representative). In this case the symmetry condition takes an especially simple form, i.e. it reduces to

$$\{f(x), \varphi(x)\} = 0 . \tag{13}$$

Such symmetries, which do not depend on, neither affect, the time, are also called *Lie-point time-independent* (or LPTI for short), and we are primarily interested in them.

Note incidentally that $\varphi \equiv f$ is an obvious solution of (13), this indeed corresponds to the "trivial" symmetry $Y \equiv X_f = f^i \partial_i$ which is just the generator of the dynamical flow: $x(t) \to x(t+t')$.

It should also be mentioned that, from a purely geometrical point of view, in which a dynamical system is just a vector field on M, it is entirely natural to consider only symmetry vector fields Y which are themselves vector fields on M, i.e. LPTI symmetries. That this is geometrically natural is reflected in the connection with the topological features of trajectories, discussed in Sect. 4 below.

Exercise 1. Given a system $\dot{x} = f(x)$, this is put in Lax form – if possible – by giving a matrix B and a matrix function $L(x)$ such that, taking into account the

4 More precisely, the evolutionary representative would be given by $X_e = (\varphi^i - \tau \dot{x}^i)\partial_i$. Thus, we are also imposing $\dot{x} = f(x)$.

time evolution of $x(t)$, these satisfy $\dot{L} = [B, L]$. Consider the system $\dot{L} = X(L)$, with the (linear) vector field X determined by $X(L) = [B, L]$, and look for symmetries of X in the form $Y = \sum_{i,j} \varphi^{ij} (\partial/\partial L_{ij})$. Characterize these, and the time evolution of the corresponding φ.

3. Constants of Motion and the Module Structure of the Symmetry Algebra

We first reintroduce some formal notation, already seen in previous chapters. We denote by $\mathcal{M}$ the Lie algebra of vector fields on M (equipped with the Lie commutator), and by $\mathcal{V}$ the algebra of functions $f : M \to \mathrm{T}M$; this is a Lie algebra if equipped with the Lie-Poisson bracket $\{.,.\}$, and there is an obvious isomorphism between $\mathcal{M}$ and $\mathcal{V}$. We also denote by $\mathcal{F}$ the algebra (actually, the ring) of scalar functions[5] $f : M \to R$. In all these sets we consider not only analytic or smooth functions (those we are interested in analyzing) but also formal power series.

We fix a given dynamical system in M, $\dot{x} = f(x)$, with $f \in \mathcal{V}$; equivalently, we fix a given vector field $X = f^i \partial_i \in \mathcal{M}$. The Lie algebra of vector fields $Y \in \mathcal{M}$ such that $[X, Y] = 0$ is denoted by $\mathcal{G}_X$. This is also characterized as the algebra of vector fields $Y = g^i(x)\partial_i$ such that $\{f, g\} = 0$, and correspondingly the Lie algebra of functions $g \in \mathcal{V}$ which satisfy $\{f, g\} = 0$ is denoted by $\mathcal{G}_f$.

The constants of motion for X are the functions $s(x) \in \mathcal{F}$ which are conserved along the flow of X, i.e. such that $X(s) = 0$ or equivalently

$$\sum_{i=1}^{n} f^i(x) \frac{\partial s(x)}{\partial x^i} = 0 . \tag{14}$$

One sees immediately these form an algebra (actually a ring), as indeed $X(s_1 + s_2) = X(s_1) + X(s_2)$. This will be denoted by $\mathcal{I} \subseteq \mathcal{F}$ (or also by $\mathcal{I}_f$ when necessary to avoid confusion).

Now, let us consider $\alpha(x) \in \mathcal{I}$ and $g(x) \in \mathcal{G}_f$, and the function $\alpha(x)g(x) \in \mathcal{V}$. This is easily seen to be again in $\mathcal{G}_f$, as

$$\{f, \alpha g\} = \alpha\{f, g\} + X(\alpha) g . \tag{15}$$

In this way, we conclude that the symmetry algebra $\mathcal{G}_f$ (equivalently, $\mathcal{G}_X$) has, beyond the structure of Lie algebra, also the structure of a *module over* $\mathcal{I}$. More precisely, we have [330]:

Proposition (Walcher). *The set $\mathcal{G}_f$ is a finitely generated module over $\mathcal{I}_f$.*

It should be emphasized that, whenever X admits a nontrivial $\mathcal{I}$, $\mathcal{G}_X$ will be infinite-dimensional as a Lie algebra, just as a consequence of the possibility of

[5] Clearly, $\mathcal{V}$ stands for "vector" and we continue to use $\mathcal{F}$ instead than $\mathcal{S}$ for the scalar function to avoid any confusion with sets of "symmetries".

multiplying any $Y \in \mathcal{G}_X$ (e.g. X itself) by a function in $\mathcal{I}$, i.e. of the module structure. Obviously, we can – and in general will – have a $\mathcal{G}$ which is infinite-dimensional as an algebra, but finite dimensional as a module.[6]

As an example of this, consider the simple system

$$\begin{aligned} \dot{x} &= \left[1 + e^{-(x^2+y^2)}\right] y \\ \dot{y} &= -\left[1 + e^{-(x^2+y^2)}\right] x \,. \end{aligned} \tag{16}$$

Clearly, $Y = x\partial_y - y\partial_x \in \mathcal{G}_X$, and $\rho = (x^2 + y^2) \in \mathcal{I}$, and thus $\mathcal{G}_X$ is infinite-dimensional as a Lie algebra, as it contains[7] the algebra $\mathcal{G}_0$ of linearly (but not functionally) independent vector fields $Y_k = \rho^k Y$. This is, however, one-dimensional as a module over $\mathcal{I}$.

Exercise 2. Consider the system

$$\begin{aligned} \dot{x}_1 &= -\,x_2 \\ \dot{x}_2 &= \;\;x_1 \;. \end{aligned}$$

Show that its Lie-point symmetry algebra $\mathcal{G}_X$, made of vector fields of the form $Y = \alpha(x)\partial_1 + \beta(x)\partial_2$ (α, β polynomials), is infinite-dimensional, and more precisely consists of all the vector fields of the form

$$Y = [a(r^2)x_1 - b(r^2)x_2]\partial_1 + [b(r^2)x_1 + a(r^2)x_2]\partial_2$$

with a, b arbitrary polynomials. Determine now $\mathcal{I}$, and check that indeed $\mathcal{G}_X$ is a module over $\mathcal{I}$. Check that any vector field $Y \in \mathcal{G}_X$ can be written in the form

$$Y = \xi(x) \sum_{k=0}^{1} c_k A^k x$$

where $(DX)(0) = Ax$ and $\xi \in \mathcal{I}_A$ is a constant of motion for the linearized equation $\dot{x} = Ax$.

4. Symmetry and Topology of Trajectories

A solution to the DS $\dot{x} = f(x)$ is a curve $\gamma \in R \times M$, i.e. a curve

$$\gamma = (t, x(t)) \tag{17}$$

in the extended phase space. However, as the ODEs representing a DS (according to our definition in Chap. II) are autonomous, it would also be entirely natural to consider just the phase space and the *trajectory* ϑ corresponding to γ, i.e. the

---

[6] The situation is similar to the one encountered when studying the invariant (scalar) and equivariant (vector) polynomials for a group representation.

[7] Actually, one could see that $\mathcal{G} = \mathcal{G}_0$.

set of points in M touched by γ. More precisely, let μ denote the projection from $R \times M$ to M, then

$$\vartheta = \{x \in M \ : \ \mu(\gamma(t)) = x, \ t \in R\} \tag{18}$$

or, with an obvious shorthand notation,

$$\vartheta \ = \ \mu(\gamma) \ . \tag{19}$$

We will not discuss the (well known) general relation between the two points of view, and the general advantages of using the standard or the extended phase space [11], and we will confine our discussion to the relation with symmetry methods.

When studying DS, a special role is played by solutions which are in some sense simple (and hopefully easier to determine than a general one) and act as organizing center. In particular, one often studies solutions of increasing order of complexity. By this we mean first of all stationary solutions, i.e. fixed points, or zeroes of the vector field X, then periodic solutions, and then n-periodic solutions.[8] Clearly, solutions $x(t)$ of these different kinds are topologically equivalent as curves $\gamma \in R \times M$, but are *topologically different* when we just look at their integral curves in M, i.e. at their trajectories ϑ. Indeed, when we speak of "increasing complexity" of solutions as above, we are actually referring to the topology, i.e. the number of fundamental cycles in the surface representing the closure of their integral curve (trajectory) in M. For the above mentioned classes of special solutions, we have respectively a point, a circle S^1, an n-torus $T^n = S^1 \times ... \times S^1$, and correspondingly, zero, one, and n cycles.

If we consider a general vector field of the form

$$Y = \tau(x,t)\partial_t + \sum_{i=1}^{n} \varphi^i(x,t)\partial_i \ , \tag{20}$$

or even a time-dependent evolutionary vector field,

$$Y = \sum_{i=1}^{n} \varphi^i(x,t)\partial_i \ , \tag{21}$$

all the solution curves γ are equivalent under these, i.e. we can a priori transform any curve into any other curve by means of a transformation generated by such a Y.

On the other hand, when we consider $Y \in \mathcal{M}$, time plays no role in the group of transformations generated by X, and we cannot transform a curve corresponding to a trajectory ϑ of a given topology into a curve corresponding

[8] Of course, other kinds of "notable" solutions exist, e.g. heterocline and homocline solutions, the ones in the stable or unstable manifolds of limit points and cycles, etc. In the present context, we will just focus on (multi)periodic ones.

to a trajectory of a different topology[9], as is seen from the group requirements, i.e. from the requirement that $e^{\lambda Y}$ must be an invertible transformation, and from the time independence of Y.

We can thus conclude that, with the notation introduced above, our discussion amounts to saying that *LPTI symmetries preserve the topology of trajectories* [144].

By the same kind of argument, we also obtain that the period of periodic solutions is preserved under LPTI transformations, and similarly for the n periods for n-periodic solutions.

In order to have a (trivial) example of this kind of situation, consider a linear system in R^{2m}

$$\dot{x} = Ax \tag{22}$$

with the matrix A written in block form, in terms of the standard two-dimensional rotation matrix

$$J = \begin{pmatrix} 0 & -1 \\ 1 & 0 \end{pmatrix} \tag{23}$$

as

$$A = \begin{pmatrix} \omega_1 J & 0 & \dots & 0 \\ 0 & \omega_2 J & \dots & 0 \\ \dots & \dots & \dots & \dots \\ 0 & 0 & \dots & \omega_m J \end{pmatrix} \tag{24}$$

For generic, i.e. independent, ω's, the LPTI symmetries of this are given, with $\rho_k = (x_{2k-1}^2 + x_{2k}^2)$ (these are the generators of $\mathcal{I}$) and α_i, β_i arbitrary smooth functions, by

$$Y_\alpha = \sum_{i=1}^m \alpha_i(\rho_1, \rho_m) \, (x_{2i-1}\partial_{2i} - x_{2i}\partial_{2i-1}) \tag{25}$$

$$Z_\beta = \sum_{i=1}^m \beta_i(\rho_1, \rho_m) \, (x_{2i-1}\partial_{2i-1} + x_{2i}\partial_{2i}) \; . \tag{26}$$

(These correspond to inhomogeneous rotations and scalings, respectively, in the various two-dimensional invariant subspaces.)

It can be checked that smoothness requires that $\alpha_i(0,...,0) = \beta_j(0,...,0) = 0$, $\forall i, j$, and this is also easily seen to imply that n-periodic solutions (with $n \geq 0$) are indeed transformed into n-periodic solutions (of the same periods).

Exercise 3. Consider the case in which some of the frequencies are multiple, i.e. there exist i, j distinct such that $\omega_i = \omega_j$ (this requires introducing a permutation group).

---

[9] It should be clear that when speaking about the "topology of a trajectory" ϑ we consider the closure $T \subseteq M$ of ϑ in M, and we mean the topology of T as a submanifold of M.

5. Time-Dependent Symmetries

In the previous section, we have seen that LPTI symmetries preserve the topology of trajectories, and we have argued that for more general transformations we are not guaranteed that this is the case.

The purpose of this section is simply to show, by means of explicit examples, that indeed by relaxing our requirement on the symmetry vector fields, e.g. considering time-dependent symmetries – or, as we will also see, symmetries that are only locally analytic – we can connect solutions whose trajectories in phase space are topologically different.

Let us consider as an example the dynamical system

$$\begin{aligned} \dot{x} &= (1-r^2)x - y \\ \dot{y} &= (1-r^2)y + x \end{aligned} \tag{27}$$

where as usual $r^2 = x^2 + y^2$. One can check easily that the vector field

$$\sigma = r^2 \, e^{-2t} \, (x\partial_x + y\partial_y) \tag{28}$$

is a symmetry of (27).

If we act with this on the periodic solution $\xi(t)$ of (27) satisfying $r(t) = 1$ (which has a circle as trajectory), applying $e^{\varepsilon\sigma}$ on $\xi(t)$ yields a solution whose trajectory is a spiral having $\xi(t)$ as asymptotic limit.

Indeed, we have

$$\begin{aligned} \frac{\partial x}{\partial \varepsilon} &= x r^2 e^{-2t} \\ \frac{\partial y}{\partial \varepsilon} &= y r^2 e^{-2t} \end{aligned} \tag{29}$$

which must be supplemented with the boundary condition $x^2+y^2 = 1$ for $\varepsilon = 0$. In this way we get

$$\frac{\partial r^2}{\partial \varepsilon} = r^4 \, e^{-2t} \, , \tag{30}$$

and we have explicitly, with obvious notation,

$$r_\varepsilon^2(t) = \left[1 - \varepsilon \, e^{-2t}\right]^{-1} . \tag{31}$$

This is a spiral, internal or external depending on the sign of ε, approaching the limit cycle $r^2(t) = 1$. ⊙

As mentioned above, the same effect, i.e. connecting solutions whose trajectories have different topologies, can also be reached by considering time-independent "symmetries", provided that we are ready to suitably enlarge the class of dynamical systems which we accept as such. In particular, we could consider vector fields which are locally, but not globally, analytic, as the following example shows.

Consider again the system (27), and denote by φ the corresponding vector field generating the time-flow. Apart from the rotation $\rho = x\partial_y - y\partial_x$, another symmetry is given by

$$\eta = (1-r^2)\,(x\partial_x + y\partial_y) \tag{32}$$

which is a combination of φ and ρ. If we multiply (32) by the (locally analytic) constant of motion

$$\kappa = \frac{r^2}{1-r^2}\,\exp\left[-2\mathrm{arctg}(y/x)\right] \ , \tag{33}$$

we get a (only locally analytic) vector field

$$\widetilde{\sigma} = r^2 \exp\left[-2\mathrm{arctg}(y/x)\right]\,(x\partial_x + y\partial_y) \tag{34}$$

which satisfies $[\widetilde{\sigma}, \varphi] = 0$. Now, on the solutions to (27) this $\widetilde{\sigma}$ coincides with the σ considered above, see (28). Thus (34) transforms limit-cycle solutions to spiralling solutions (or viceversa), i.e. it connects solutions which have trajectories with different topologies. ⊙

It should be remarked that these "local" symmetries[10] are quite common. Indeed, from the fundamental theorem on rectification of flow [11] we know that around any non-singular point for a flow in R^n we have, beyond the flow itself, $(n-1)$ local symmetries and $(n-1)$ local constants of motion.

It can be useful to note that, if one finds n functionally independent constants of motion $s_i = s_i(x,t)$ of the DS (14) (here one must also consider constants of motion possibly dependent on time), then one can introduce the following quantities $p_{jk}(x,t)$ by means of the linear system (in fact, $\det(\partial s_j \,/\, \partial x_i) \neq 0$ in some open domain)

$$\sum_j p_{jk}\,\frac{\partial s_j}{\partial x_i} = \delta_{ik} \qquad (i,j,k = 1,\dots,n) \ .$$

Then one can verify that the vector fields

$$Y_j = \sum_k p_{jk}\frac{\partial}{\partial x_k}$$

provide n independent symmetries of the DS $\dot{x} = f$ [266].

Let us emphasize that in the following we will only be concerned with LPTI symmetries that are expressed as power series (either formal or possibly convergent in some neighborhood of $x_0 = 0$).

Exercise 4. Consider the system

$$\dot{x} = -\,r^2 y$$
$$\dot{y} = \ r^2 x \ .$$

All of its solutions are periodic, and the period depends on r. Show that

$$\sigma = (x\partial_x + y\partial_y) \, + \, 2\mathrm{arctg}(y/x)\,(y\partial_x - x\partial_y)$$

connects solutions with different periods.

[10]They should not be confused with a local group of symmetries: here we are considering vector fields that are well defined only locally.

6. Orbital Symmetries

In the previous sections, we focused on the trajectory θ rather than on the full solution γ (Sect.4). However, we considered the effect of proper Lie-point symmetry on these. By "proper" symmetries, we mean vector fields which transform the solution γ into another (possibly, the same) solution γ'.

In recent works, Walcher [332,333] considered *orbital symmetries*, mainly focusing on the case where these leave the (orbit of the) solution invariant[11], but do not necessarily leave the solution itself invariant. In this section, we just want to point out the existence of this promising theory, following [332]. For a more complete discussion, see [333].

Given two dynamical systems on M identified by X_f and X_g, i.e.

$$\dot{x} = f(x) \quad \text{and} \quad \dot{x} = g(x) \ , \tag{35}$$

we say that $\Phi : M \to M$ is a solution-preserving map from f to g if $\Phi \otimes I$ maps solutions $(x(t), t)$ of $\dot{x} = X_f(x)$ to solutions $(y(t), t)$ of $\dot{x} = X_g(x)$. If $g(x) = \mu(x) f(x)$ with $\mu(x)$ a scalar function $\mu : M \to R$, and Φ is invertible, we say that Φ is an *orbital symmetry* of f (a proper orbital symmetry if $\mu(x) \not\equiv 1$). The group G of maps $\Phi : M \to M$ is an orbital symmetry group for f if all the $\Phi \in G$ are orbital symmetries of f.

The above definition of orbital symmetries is justified by the fact that the vector fields X_f and $X_{\mu f}$ have different solutions but the same trajectories (orbits), and conversely two dynamical systems which have locally the same solution trajectories differ only by a scalar factor $\mu(x)$ which is, moreover, the quotient of two analytic functions [332].

Orbital symmetries are more abundant than proper symmetries (indeed, any proper symmetry is also an orbital symmetry, but the converse is not true). Correspondingly, the reduction procedure by orbital symmetries is more complicated than the one by proper symmetries [332,333]. As far as we know, reduction by orbital symmetries has been considered only for finite linear groups of orbital symmetries [332].

In this case, orbital symmetry groups can be characterized as follows [332,333]: A finite linear group G is an orbital symmetry group of $\dot{x} = f(x)$ if and only if there is a homomorphism $\chi : G \to C^*$ such that

$$\Phi \cdot f \cdot \Phi^{-1} \ = \ \chi(\Phi) f \tag{36}$$

for all the $\Phi \in G$. When $\chi(\Phi) = 1$, we have a proper symmetry group.

We refer to [333] for further details and applications.

Exercise 5. Is every vector field that leaves invariant the trajectories of solutions to a differential equation $\dot{x} = f(x)$ an orbital symmetry ? What if we consider a time-dependent vector fields ?

---

[11]Instead of transforming the orbit – i.e. in our terminology the trajectory – of one solution into that of a different solution.

7. Approximate Symmetries

As we have been seeing, the knowledge of exact symmetries of an ordinary differential equation allows us to reduce it, and sometimes to completely solve it. Thus, it helps in determining exact solutions.

However, in many cases one would be satisfied with approximate rather than exact solutions. It turns out, as we will briefly discuss in this section, that in this case *approximate symmetries* (to be introduced below) are as good as exact ones. Moreover, it happens that these can be determined perturbatively, which is of not little help.

We will, as usual, consider autonomous dynamical systems written as

$$\dot{x} = f(x) \tag{37}$$

with $x \in M \subseteq \mathbf{R}^n$, $f : M \to TM$, and we will assume that $f(0) = 0$. We also write $X = f^i(x)\partial_i$ for the associated vector field.

We recall that under a coordinate transformation generated by the vector field $Y = s^i(x)\partial_x$, i.e.

$$x \to \widetilde{x} = x + \varepsilon s(x) + o(\varepsilon) \ , \tag{38}$$

this is transformed to a new system

$$\widetilde{f}(x) := f(x) + \varepsilon\{f, s\} + o(\varepsilon) \ , \tag{39}$$

where we have introduced the Lie-Poisson bracket

$$\{\varphi, \psi\} := (\varphi \cdot \nabla)\psi - (\psi \cdot \nabla)\varphi \ . \tag{40}$$

The LPTI symmetries of (37) are therefore given by vector fields Y with s solution of

$$\{f, s\} = 0 \ . \tag{41}$$

This is just the commutation condition

$$[X, Y] \ = \ 0 \tag{42}$$

expressed in terms of the functions f, s.

In order to determine explicitly Y, we should solve (41). we now want to solve this perturbatively. Thus, we expand f and s around $x_0 = 0$ (note that necessarily $s(0) = 0$); i.e. we write

$$f(x) = \sum_{m=0}^{\infty} f_m(x) \tag{43}$$

$$s(x) = \sum_{m=0}^{\infty} s_m(x) \ , \tag{44}$$

where f_m, s_m are homogeneous polynomials of degree $m + 1$. Writing (41) in terms of these gives a series of equations

$$\sum_{j=0}^{k} \{f_j, s_{k-j}\} = 0 \qquad k = 0, 1, 2, ... \tag{45}$$

Note that if we have solved the equations in this sequence for $k \le k_0 - 1$, i.e. have determined $s_0, ..., s_{k_0-1}$, then the k_0 equation reads

$$\{f_0, s_{k_0}\} = -\sum_{j=1}^{k_0} \{f_j, s_{k_0-j}\} := F_{k_0}(x) \tag{46}$$

where the r.h.s. is a known function of $f_0, ..., f_{k_0-1}$ and $s_0, ..., s_{k_0-1}$.

Remark 1. Notice that $\{f_0, .\}$ is the *homological operator* associated with (the linear part of) f of normal form theory (Chaps. II and IV). With $f_0(x) = Ax$, we also write[12]

$$\{f_0, .\} := L_A(.) \tag{47}$$

so that (46) is an equation of the form

$$L_A(s_k) = F_k(x) \tag{48}$$

that can be solved if and only if $F_k(x) \in \mathrm{Ran}(L_A)$. Moreover, s_k is determined up to a function in $\mathrm{Ker}(L_A)$.

When (41), i.e. (45), is satisfied not for all k, but only for $k \le K$, we say that Y is an **approximate symmetry** of order K for X.

We can also write, for given f (i.e. X) and s (i.e. Y),

$$[X, Y] = Z = r^i \partial_i = \left(\sum_k r_k^i \right) \partial_i \tag{49}$$

which is equivalently written as

$$\{f, s\} := r \;\; ; \;\; r(x) = \sum_{m=0}^{\infty} r_m(x) \; . \tag{50}$$

With this notation, we have the following:

Definition 1. *If in* (49),(50) *we have* $r_m = 0 \; \forall m \le k$, *then* $Y = s^i \partial_i$ *is an approximate symmetry of order* k *for (1).*

The set of approximate symmetries of order k of the vector field X will be denoted by $\mathcal{G}_X^{(k)}$; obviously, $\mathcal{G}_X^{(k+1)} \subseteq \mathcal{G}_X^{(k)}$.

Lemma 1. *The set* $\mathcal{G}_X^{(k)}$ *is a a Lie algebra.*

Proof. Let Y_1, Y_2 be in $\mathcal{G}_X^{(k)}$. We consider

[12]We repeat here some statements already presented – and studied – in previous chapters for the sake of completeness of the present chapter.

$$[Y_1, Y_2] \equiv Z = s^i(x)\partial_i \ ; \quad s(x) = \sum_{m=0}^{\infty} s_m(x) \tag{51}$$

and want to show that σ is also an approximate symmetry. The commutator

$$[X, Z] \equiv W = w^i(x)\partial_i \ ; \quad w(x) = \sum_{m=0}^{\infty} w_m(x) \tag{52}$$

is also written, using the Jacobi identity, as

$$[Y_1, [X, Y_2]] - [Y_2, [X, Y_1]] \equiv [Y_1, R_2] - [Y_2, R_1] \ . \tag{53}$$

Now, both R_1 and R_2 have only terms of order greater than k, so that $w_m(x) = 0$ for all the $m \le k$; that is, $Z \in \mathcal{G}_X^{(k)}$. ⊙

Actually, as was already the case for the symmetry algebra $\mathcal{G}_X$, the approximate symmetry algebra of order k also has the structure of a module.

Let us consider a scalar function $\zeta(x) \in R$, which is expanded as

$$\zeta(x) = \sum_{m=0}^{\infty} z_m(x) \ . \tag{54}$$

By acting on this with X we get

$$X(\zeta) = \sum_{m=0}^{\infty} \sum_{k=0}^{\infty} (f_k \cdot \nabla) z_{(m-k)}(x) \equiv \sum_{m=0}^{\infty} a_m(x) \ . \tag{55}$$

Definition 2. *If in* (55) *we have* $a_m(x) \equiv 0$ *for all the* $m \le k$, *we say that* ζ *is an approximate constant of motion of order* k *for* X.

We denote by $I_X^{(k)}$ the set of approximate constants of motion of order k for X. Clearly, $I_X^{(k)}$ is an (abelian) algebra under the standard product of functions; moreover we have $I_X^{(k+1)} \subseteq I_X^{(k)}$.

Lemma 2. *The set* $\mathcal{G}_X^{(k)}$ *has, beyond the structure of Lie algebra, the structure of a module over the algebra* $I_X^{(k)}$.

Proof. This follows immediately from (55). ⊙

Remark 2. The determination of approximate symmetries and approximate constants of motion is intimately related to the construction of normal forms. This relation is discussed in [94] for general systems. For the special case of Hamiltonian systems, see [159] or Chapter IV.

Remark 3. In many cases[13] the study of nonlinear systems goes through consideration of a suitable truncation of the nonlinear equation, followed by a study of the relation between the dynamics of the full system and that of the truncated one. In this respect, approximate symmetries (of suitable order) are as relevant as the exact ones, as they play the same role in the study of the truncated system [94].

---

[13]In particular, but not only, when using the normal forms approach.

Appendix. On the Module Structure

In Sect. 3 we have shown the module structure of the symmetry algebra $\mathcal{G}_X$ for a vector field $X = f^i(x)\partial_i$; a more detailed and complete discussion of this structure will be given in Chap. IV.

The present short discussion is aimed at preventing a few common misunderstandings concerning the structure of $\mathcal{G}_X$, and related questions. Indeed, one could be led to think that the form given in Exercise 2 is the general one for symmetry vector fields, i.e. that the above statement holds for general X and A [with $(Df)(0) = A$ and A diagonalizable]. It should be stressed that, for general systems, this is *not* true.

The correct relation between linearized systems, their symmetries and constants of motions, and symmetries and constants of motion for the full system, will be discussed in Chap. IV (and is also given in [121]). Here we just want to convince the reader, by explicit example, that this cannot be as simple as the one satisfied for the simple rotation vector field considered above.

Consider the system $\dot{x} = f(x)$ $(x \in R^2)$ with [J is the same as in (23)]

$$f(x) = [\alpha(r^2)I + \beta(r^2)J]x \ . \tag{A.1}$$

It is easy to see that for general α, β, there are symmetries of this which cannot be written in the form $Y = G^i(x)\partial_i$ with

$$G(x) = \xi(x)\sum_{k=0}^{1} c_k A^k x \tag{A.2}$$

where $(Df)(0) = A$ and $\xi \in \mathcal{I}_A$ is a constant of motion for the linearized equation $\dot{x} = Ax$.

Indeed, it suffices to consider the case where $\alpha(0) = 1$ and $\beta(0) = 0$, and to notice that $X \in \mathcal{G}_X$ by definition [$X = X_f$ is the vector field corresponding to (A.1)]: now $A = I$, and of course $A^0 = I$, but r is *not* a constant of motion of $\dot{x} = Ax$, and $J \neq A^n$ for all n, so that there is no way to write $f(x)$ in the form (A.2). A similar reasoning would apply also to the case where $\beta(0) \neq 0$ (so that we have distinct eigenvalues, see below).

It should be stressed, again as a warning against possible errors, that the unfolding of the normal form of vector fields X in R^N with given linear part A is *not* – for general systems, i.e. general A – of the form

$$\widetilde{F}(x) = Ax + \sum_{m,k} c_{m,k}\xi_m(x)A^k x \ , \tag{A.3}$$

where $\ell = 0, 1, ..., N-1$ and the ξ_m are a basis for I_A, the ring of constants of motion under $\dot{x} = Ax$.

To see that (A.3) is not correct, it will be enough to consider the case

$$A = \begin{pmatrix} 1 & 0 & 0 \\ 0 & 1 & 0 \\ 0 & 0 & -2 \end{pmatrix} . \tag{A.4}$$

Note that in this case we can write the normal form in a way similar to (A.3), provided that we consider not only the A^k, but all the matrices which commute with A (again, this will be discussed in full in Chap. IV, see also [330]). Note that in [204], formula (A.3) is given, but the author specifies that A is assumed to be diagonalizable, and to have only eigenvalues with multiplicity one, in which case the A^k provide a basis for the linear space of matrices commuting with A.

It seems to us that, at least in some cases, the origin of the error mentioned above can be traced to a rather simple misunderstanding: if $X = f^i(x)\partial_i$, and $Y = S^i\partial_i$ commute with X, then

$$\frac{\mathrm{d}S^i}{\mathrm{d}t} = \frac{\partial f^i}{\partial x^j} S^j \,. \tag{A.5}$$

However, if in the same setting we consider $(Df)(0) = A$ then, in general, Y (i.e. its components S^i) does not evolve according to the linearized equations:

$$\frac{\mathrm{d}S^i}{\mathrm{d}t} \neq A\,S^j \,. \tag{A.6}$$

We also stress that, in general, not only are the symmetries not solutions of the linearized equation $\dot{x} = Ax$, but also they are not even symmetries for it.

Note that the situation is, in several regards, different if X is in normal form, as discussed in Chap. IV.

IV. Normal Forms and Symmetries for Dynamical Systems

In this chapter, we consider a smooth dynamical system (DS) on a manifold M, which we take, for ease of notation, as being embedded[1] in $\mathbf{R}^n$, with a zero at a point x_0 which we take, again for ease of notation, to be at the origin of $\mathbf{R}^n$. We have then

$$\dot{x} = f(x) \qquad f : M \to \mathrm{T}M \ , \ f(0) = 0 \ , \tag{1}$$

or, equivalently, a tangent vector field (VF) on M,

$$X = \sum_{i=1}^{n} f^i(x) \partial_i \qquad \left(\partial_i \equiv \frac{\partial}{\partial x^i} \right) \ , \tag{2}$$

and we are interested in the normal forms classification of such DS, or equivalently of such VF.

More precisely, we are interested in how the symmetry properties of (1) and (2) are reflected in their normal form (NF), and conversely how the knowledge of such symmetry properties can help us in determining the normal form.

In the following, it will be convenient to work in the category of (possibly formal) power series rather than smooth or analytic functions.

1. Perturbative Expansion of Determining Equations

Let us consider (1) again, and expand f in a perturbative series around $x_0 = 0$. We write, as usual,

$$f(x) = \sum_{k=1}^{\infty} F_k(x) \tag{3}$$

where the F_k are homogeneous of degree k, i.e.

$$F_k(ax) = a^k F_k(x) \qquad (a \in \mathbf{R}) \ . \tag{4}$$

In order to determine the symmetries of (1), we should consider a generic smooth vector field on $M \times \mathbf{R}$,

---

[1] Actually, all the considerations will be local around the fixed point x_0, as always in normal form theory. Thus, in practice we work only in $\mathrm{T}_{x_0}M$, and can take $M = \mathbf{R}^n$ with no loss of generality.

$$V = v^0(x,t)\partial_t + \sum_{i=1}^n v^i(x,t)\partial_i \ . \tag{5}$$

However this is too general for our purposes, as it does not take into account the distinguished nature of time, and could map a DS into an equation of different nature. We are primarily interested in vector fields acting on M alone, and actually it turns out [95,146] that one is mainly interested in time-independent VF on M, i.e.[2]

$$Y = \sum_{i=1}^n s^i(x)\partial_i \ . \tag{6}$$

It will be convenient to have a notation for the smooth[3] vector fields on M, so we denote these by $\mathcal{M}$. Clearly, there is a one to one correspondence between vector fields defined on M and the set $\mathcal{V}$ of smooth functions from M to $\mathrm{T}_x M$, which can also be seen as sections of the bundle $\mathrm{T}M$.

Now, let us consider the DS (1), where $X \in \mathcal{M}$, and look for the determining equation identifying its symmetries.[4] According to the general procedure, we should consider the prolongation of the generic vector field V, which will be $V^{(1)} = V + \sum_i w^i(\dot{x},x,t)(\partial/\partial\dot{x}^i)$, with w^i determined by the prolongation formula, and apply it to (1), which gives

$$w^i(\dot{x},x,t) = \sum_j v^j \frac{\partial f^i}{\partial x^j} \ . \tag{7}$$

Restricting V to be of the form (6), i.e. considering instead $Y \in \mathcal{M}$, we get for w^i just $(\partial s^i/\partial x^j)\dot{x}^j$, and restriction to the solution manifold of (1), i.e. consideration of the constraint $\dot{x} = f(x)$ given by (1) itself, yields

$$f^j(x)\frac{\partial s^i(x)}{\partial x^j} - s^j(x)\frac{\partial f^i(x)}{\partial x^j} \equiv (f\cdot\nabla)s - (s\cdot\nabla)f = 0 \ . \tag{8}$$

In Chap. II this has also been written in terms of the antisymmetric bracket $\{.,.\}$, defined as

$$\{f,g\} \equiv (f\cdot\nabla)g - (g\cdot\nabla f) \ , \tag{9}$$

and we have that $Y = s^i\partial_i \in \mathcal{M}$ is a symmetry of $X = f^i\partial_i \in \mathcal{M}$ if and only if the two equivalent conditions below are realized:

$$\{f,s\} = 0 \qquad ; \qquad [X,Y] = 0 \ . \tag{10}$$

We denote the set of Y satisfying (10) by $\mathcal{G}_X$, and – since no confusion will be possible here – we will just write $\mathcal{G}$. It is obvious that $\mathcal{G} \subset \mathcal{M}$, equipped with

---

[2] Note that in this chapter we will consistently use X for X_f and Y for X_s, in order to simplify the notation.

[3] By smooth we mean, here and elsewhere, C^∞; however, we will soon go on to actually consider vector fields represented by power series.

[4] We briefly repeat the discussion presented in Chap. II.

the usual commutator of vector fields, is a Lie algebra. It is also easy to see that, as $f(0) = 0$, $Y \in \mathcal{G}$ implies that $s(0) = 0$.

We can now expand Y, i.e. the $s(x)$, in a perturbative series around $x_0 = 0$ in the same way as we did for X. That is, we write

$$s(x) = \sum_{k=1}^{\infty} S_k(x) \qquad (S_k(ax) = a^k S_k(x)) \ . \tag{11}$$

We can now plug (11) and (3) into the determining equations (10). In this way we get, after rearranging the terms,

$$\sum_{k=1}^{\infty} \sum_{m=1}^{k-1} \{F_m, S_{k-m}\} = 0 \ , \tag{12}$$

and for this analytic expression to vanish, the different homogeneous terms of degree k must vanish separately. Thus, we have a hierarchy (in a sense to be explained in a moment) of equations

$$\sum_{m=1}^{k-1} \{F_m, S_{k-m}\} = 0 \qquad k = 2, 3, \dots \ . \tag{13}$$

It is convenient to isolate the terms containing linear factors, i.e. to rewrite (13) in the form

$$\{F_1, S_{k-1}\} - \{S_1, F_{k-1}\} = - \sum_{m=2}^{k-2} \{F_m, S_{k-m}\} \equiv \psi_k \ , \tag{13'}$$

where we have used the antisymmetry of $\{.,.\}$, and $\psi_2 = \psi_3 = 0$.

Exercise 1. Consider the vector field in $\mathbf{R}^2$ given by $X = (x + xy - x^2 + x^2 y)\partial_x + (y + y^2 - x^2 y + xy^2)\partial_y$. Write down the determining equations for a generic vector field $Y = \alpha(x,y)\partial_x + \beta(x,y)\partial_y \in \mathcal{M}$ to be in $\mathcal{G}_X$, in the form (12).

Exercise 2. Consider the linear vector field in $\mathbf{R}^n$ associated with a matrix A, $X = A_{ij} x^j \partial_i$; let Φ be an invertible diffeomorphism of $\mathbf{R}^n$, and let z be the coordinates obtained as $x = \Phi(z)$. Write down the X vector field in the z coordinates. Determine the vector fields $Y = s^i(z)(\partial/\partial z^i)$ commuting with X.

2. Recursive Determination of Symmetries

Let us now consider the problem of concretely solving the determining equation (12). As the perturbative series expansion suggests, we can proceed order by order and use the hierarchy of equations (13), i.e. start with consideration of the equation for $k = 1$, then tackle $k = 2$, and so on. The reason for proceeding in such a sequential way is that when we consider the k-th equation, we have the term S_k to be determined, but the terms S_m for $m \le (k-1)$ will already have been determined, and these are those appearing in the ψ_k. Namely, by proceeding in this way we are always reduced to considering an equation of the form

$$F_1^j \left(\frac{\partial S_k^i}{\partial x^j} \right) = S_k^j(x) \left(\frac{\partial F_1^i}{\partial x^j} \right) + \psi_k^j(x) \tag{14}$$

with the ψ_k^j known functions of x.

Notice also that F_1 is just the (known) linear part of X. If we write it in matrix form as

$$F_1^i(x) = A_{ij} x^j \ , \tag{15}$$

then (14) reads simply

$$[A_{ij} x^j] \left(\frac{\partial S_k^i}{\partial x^j} \right) = A_{ij} S_k^j + \psi_k^j(x) \ . \tag{16}$$

Thus, solving the determining equations in such a recursive way only requires to solve at each stage a non-homogeneous system of linear PDEs for the S_k.

Let us follow explicitly this solution for the first steps. For $k = 2$, we only have to require that

$$\{F_1, S_1\} = 0 \ . \tag{17}$$

Using (15) for F_1 and writing similarly

$$S_1^i(x) = B_{ij} x^j \tag{18}$$

for S_1, we get

$$\{F_1, S_1\} \equiv \{Ax, Bx\} = -[A, B]_{ij} \, x^j \tag{19}$$

and therefore at this stage we only have to determine the matrices B commuting with a given matrix A.

For further reference, we introduce the notation X_A for the linear vector field associated with the matrix A, i.e.

$$X_A = (A_{ij} x^j) \frac{\partial}{\partial x^i} \ . \tag{20}$$

and we notice that in this notation (19) yields the

Lemma 1. *For any two matrices A, B, the commutator of the associated linear vector fields is the linear vector field associated with the commutator of the matrices with a minus sign, i.e.* $[X_A, X_B] = -X_{[A,B]}$.

Let us go back to the solution of the determining equations. For $k = 3$, we have

$$\{Ax, S_2\} = \{Bx, F_2\} \ . \tag{21}$$

so that we have (16) with

$$\psi_1^i = \{Bx, F_2\}^i \ . \tag{22}$$

At this point, it is worth pausing a moment to remark that the operator $\{Ax, .\}$ is just the *homological operator* associated with A which one considers in normal form theory (Chap. II). Similarly, $\{Bx, .\}$ is the homological operator associated with B. We will write in the following L_M for the homological operator associated with a matrix M, and we will also use the shorthand notation

$$\mathcal{A} = L_A = \{Ax, .\} \quad , \quad \mathcal{B} = L_B = \{Bx, .\} \ . \tag{23}$$

With this notation, (21) is just $\mathcal{A}(S_2) = \mathcal{B}(F_2)$, and more generally we get at stage k

$$\mathcal{A}(S_k) = \mathcal{B}(F_k) + \psi_k \ . \tag{24}$$

As is clear from the above discussion, the homological operators do play a crucial role also in the determination of symmetries.

It should be remarked explicitly that, as a simple consequence of the Jacobi identity, we have that:

Lemma 2. *For any matrices A, B, the commutator of the associated homological operators is equal to the homological operator associated with the commutator with a minus sign, i.e.* $[L_A, L_B] = L_{[B,A]} = -L_{[A,B]}$.

Proof. Indeed, $[L_A, L_B]f = \{Ax, \{Bx, f\}\} - \{Bx, \{Ax, f\}\} = \{\{Ax, Bx\}, f\}$ and, using (19), this is just $\{[A, B]x, f\} \equiv -L_{[A,B]}(f)$. ⊙

The fact that we can proceed recursively, and only deal with linear PDEs, does not mean, obviously, that we are guaranteed to find solutions at any stage k. At $k = 1$ we always have at least the solutions given by $B = I$ and by $B = A^k$, but for $k \geq 2$ we do not, in general, have solutions to the determining equations apart from $Y = cX$. (Notice that for some systems we could have symmetries in different classes, e.g. corresponding to non analytic or singular functions.)

Note also that, as the relevant equations are not homogeneous, even $S_k = 0$ is not, in general, an acceptable solution at stage k. This is quite natural if one thinks that the choice $B = I$ is always acceptable at $k = 1$. Choosing $S_k = 0$ at all the following stages would leave us with the dilation vector field $Y = x^i \partial_i$, which is a symmetry only for X linear.

Exercise 3. Consider the vector fields $X = f^i(x)\partial_i$ and $Y = s^i(x)\partial_i$ in $\mathbf{R}^2$ given by

$$f(x) = \begin{pmatrix} -x_2 \\ x_1 \end{pmatrix} \quad ; \quad s(x) = \begin{pmatrix} ax_1 + b(x_1^2 + x_2^2)x_2 \\ ax_2 - b(x_1^2 + x_2^2)x_1 \end{pmatrix} \ .$$

Do these satisfy the equations (24) ?

Exercise 4. Consider the vector field $X = f^i(x)\partial_i$ in $\mathbf{R}^2$ given by $f^1(x) = a[1+b(x_1^2+x_2^2)]x - c[1+d(x_1^2+x_2^2)]$, $f^2(x) = a[1+b(x_1^2+x_2^2)]y + c[1+d(x_1^2+x_2^2)]$. Determine recursively its symmetries by solving equations (24) up to order three.

3. Approximate Symmetries

It is natural, and also useful, to consider the *approximate symmetries* that were introduced in Chap. III. Consider the vector fields X and Y such that

$$[X,Y] \;=\; Z \;=\; u^i(x)\partial_i \;. \tag{25}$$

We recall that Y is an approximate symmetry of order m for X if in the expansion of u in power series,

$$u^i(x) \;=\; \sum_{k=1}^{\infty} U_k^i(x) \qquad \left(U_k^i(ax) = a^k U_k^i(x)\right) \;, \tag{26}$$

all the terms of order smaller or equal to m are zero,

$$U_k^i(x) \equiv 0 \quad \forall k \le m \;. \tag{27}$$

We denote the set of approximate symmetries of X of order m by $\mathcal{G}_X^{(m)}$ or, when no confusion is possible, by $\mathcal{G}^{(m)}$. It is easy to check that $\mathcal{G}^{(m)}$, with the usual commutator of vector fields, is a Lie algebra. It is also easy to check that the chain of $\mathcal{G}^{(m)}$ is a filtration of Lie algebras. We have:

Lemma 3. *For any vector field $X \in \mathcal{M}$, the chain of the approximate symmetry algebras satisfies $\mathcal{G}_X^{(k+1)} \subseteq \mathcal{G}_X^{(k)}$. The proper symmetry algebra $\mathcal{G}_X$ corresponds to the limit of $\mathcal{G}_X^{(k)}$ for $k \to \infty$.*

It is appropriate to consider, together with approximate symmetries, also *approximate constants of motion.* These are defined in a way similar to that for approximate symmetries.

Consider the ring $\mathcal{F}$ of analytic functions from M to $\mathbf{R}$. Let $\rho \in \mathcal{F}$ and consider the power series expansion of $X(\rho)$ around $x_0 = 0$:

$$X(\rho) \;=\; r(x) \;=\; \sum_{k=1}^{\infty} R_k(x) \;. \tag{28}$$

We say that ρ is an approximate constant of motion of order m for X if all the R_k are zero up to order m,

$$R_k(x) \equiv 0 \qquad \forall k \le m \;. \tag{29}$$

We denote the set of approximate constants of motion of order m for X by $\mathcal{I}_X^{(m)}$. It is easy to check that $\mathcal{I}_X^{(m)}$, equipped with the usual sum and product of functions, is a ring. In particular, it is an abelian algebra under the sum, $[f+g](x) = f(x) + g(x)$. We have then:

Lemma 4. *For any vector field $X \in \mathcal{M}$, the chain of the approximate constants of motion rings satisfies $\mathcal{I}_X^{(k+1)} \subseteq \mathcal{I}_X^{(k)}$; the proper constants of motion ring $\mathcal{I}_X$ corresponds to the limit of $\mathcal{I}_X^{(k)}$ for $k \to \infty$.*

We have seen in Chap. III that the set $\mathcal{G}$ has, beyond the structure of a Lie algebra, also the structure of a module over the algebra of constants of motion $\mathcal{I}$. The same happens for the sets $\mathcal{G}^{(k)}$, i.e. we have:

Lemma 5. *The set $\mathcal{G}^{(k)}$ has, beyond the structure of a Lie algebra, also the structure of a module over the algebra $\mathcal{I}^{(k)}$ of approximate constants of motion of the same order.*

We would like to emphasize that when performing actual perturbative computations, approximate symmetries and/or constants of motion are as good as exact ones. Indeed any actual computation is only up to some finite order, and at each order we can use the approximate symmetries of that order as well as the exact symmetries.

It turns out that the determination of symmetries of a DS and its transformation into normal form are not only closely related by the role of the homological operator, but actually are better accomplished if performed in parallel.

Let us emphasize that these normal form and perturbation techniques are useful tools in the study of many physical problems and fields. We mention here: [35,37,115,184,311,313] for applications concerning the physics of elementary particle accelerators, [36] for problems in magnetically confined plasmas and beam dynamics, [267] for applications to the Lorenz model, and [10] for a recent and relevant application to stochastic equations. Several other applications will be mentioned in the next chapters, being especially related to the case of Hamiltonian systems, both in classical and in quantum mechanics.

We also point out that several algorithms and computer packages for obtaining normal forms have been developed [198,296,199,276]; see also [98] for a discussion of the use of normal forms for dynamical systems from the point of view of computational algorithms.

Several extensions (which will not be discussed here) of normal forms techniques have been proposed, using, e.g., changes of coordinates of different type: for instance introducing differentiable (but non-analytic) transformations [306,307], nonlocal changes of coordinates [226], and also a time-dependent procedure [310]. See also [71,72,206,336,346] for some other recent contributions.

Exercise 5. Reinterpret your answer to Exercise 4 in view of the results given in this section, and in particular of Lemma 5.

Exercise 6. Prove Lemma 5.

4. Symmetry Characterization of Poincaré-Dulac Normal Forms

Let us now consider the *Poincaré-Dulac normal form* for the dynamical system (1), introduced in Chap. II. We recall that this corresponds to changing coordinates, via a formal C^∞ change of coordinates $x = x(y)$, so that in the new coordinates X reads

$$X = g^i(y)\,\frac{\partial}{\partial y^i} \tag{30}$$

where the linear terms in g are the same as in the f, and the nonlinear terms in g are in the kernel of the relevant homological operator (associated with the linear part); namely,

$$g^i(y) = \sum_{k=1}^{\infty} G^i_k(y)\,, \tag{31}$$

$$G^i_1(y) = A_{ij}y^j \quad , \quad G_k \in \mathrm{Ker}\,(L_{A^+}) \quad \text{for } k > 1\,. \tag{32}$$

It should be recalled that $h \in \mathrm{Ker}\,(L_A)$ means that the vector field associated with h (which we denote by X_h) commutes with the linear vector field X_A associated with A, i.e.

$$h \in \mathrm{Ker}\,(L_A) \iff \{Ax, h(x)\} = 0 \iff [X_A, X_h] = 0\,. \tag{33}$$

Let us summarize the definition given in Chapter II, Sect.5 and the above remarks in the following form:

Lemma 6. *A vector field $X = \left(A_{ij}y^j + h^i(y)\right)\partial_i$, where the h are nonlinear functions, is in normal form if and only if its nonlinear part $X_h = h^i(y)\partial_i$ commutes with the vector field $X_{A^+} = \left[A^+_{ij}y^j\right]\partial_i$ associated with the adjoint of its linear part (i.e. if and only if $h \in \mathrm{Ker}\,(L_{A^+})$).*

We will first consider (in this and the following sections) the case where the matrix A, corresponding to the linear part of X in both the original and the normal form coordinates, commutes with its adjoint; i.e. we make the

Assumption. *The matrix A is normal:* $[A, A^+] = 0$.

We take this as a blanket hypothesis until Sect. 10 of this chapter, where we will consider the extension of the results obtained under this assumption to the generic case.

If A is normal, then $[L_A, L_{A^+}] = 0$ due to Lemma 2; this implies in particular that we have

Lemma 7. *If A is normal, then* $\mathrm{Ker}\,(L_A) = \mathrm{Ker}\,(L_{A^+})$.

Proof. Denote by $\{\lambda_1, ..., \lambda_n\}$ the eigenvalues of A; then the space $\mathrm{Ker}\,(L_A)$ is spanned by the functions $y_{\mu,s}$ (defined in components by $y^i_{\mu,s} = x^\mu \delta_{i,s}$, with $\mu = (\mu_1, ..., \mu_n)$ a multiindex, the μ_i being non-negative integers) which satisfy the resonance condition $\sum_i \lambda_i \mu_i = \lambda_s$. As the eigenvalues of A^+ are $\{\lambda^*_1, ..., \lambda^*_n\}$, the (μ, s) for which the resonance relations are satisfied are the same, and thus $\mathrm{Ker}\,(L_A)$ and $\mathrm{Ker}\,(L_{A^+})$ are spanned by the same vectors of an orthonormal basis. ⊙

It is important to recall that, as shown in Chap. II, with the standard scalar product in $\mathcal{V}$ – i.e; in $\mathcal{M}$ – we have $(L_A)^+ = L_{A^+}$. It is also important, although trivial, to notice that

Lemma 8. *If A is a normal matrix,* $\mathrm{Ker}(L_A^2) = \mathrm{Ker}(L_A)$.

Proof. If A is normal, then L_A is a normal operator (under the standard scalar product), and $\mathrm{Ker}(L_A) \cap \mathrm{Ran}(L_A) = \{0\}$; but, $f \in \mathrm{Ker}(L_A^2) \Rightarrow (L_A(f)) \in \mathrm{Ker}(L_A)$ and therefore $L_A(f) = 0$. ⊙

This discussion leads to a natural characterization of Poincaré-Dulac normal forms in terms of symmetry. Namely, we have the

Proposition. *If A is a normal matrix, then X is in normal form if and only if X_A is a symmetry of X, i.e. if and only if it commutes with its linear part*

$$[X, X_A] = 0 \qquad \text{or, equivalently} \qquad \{g(y), Ay\} = 0 \; . \tag{34}$$

We also say that the terms in $\mathrm{Ker}(L_A)$ are *resonant*. Thus, X is in normal form if and only if all its terms are resonant.[5]

Exercise 7. Consider (2×2) real matrices; characterize the normal matrices among these.

Exercise 8. Assume that the matrix A is semisimple. Is it always true that in this case $[A, A^+] = 0$? Is it always possible to change A into a normal matrix by a linear transformation ?

5. Nonlinear Symmetries and Normal Forms

We want to now consider the case where the DS (1), or equivalently the vector field X, admits a symmetry Y (the case of an n-dimensional symmetry algebra will be considered later on). We will use the notation introduced above for the expression of X, Y in the x coordinates, and denote by y the normal form coordinates. We have, therefore,

$$\begin{aligned} X &= f^i(x) \frac{\partial}{\partial x^i} &= g^i(y) \frac{\partial}{\partial y^i} \\ Y &= s^i(x) \frac{\partial}{\partial x^i} &= r^i(y) \frac{\partial}{\partial y^i} \end{aligned} \tag{35}$$

and similarly for the power series expansions of f, g, s, r in terms homogeneous of degree $k = 1, 2, \ldots$. We will denote the matrices associated with the linear parts of X, Y by, respectively, A and B:

$$(Df)(0) = (Dg)(0) = A \quad , \quad (Ds)(0) = (Dr)(0) = B \; . \tag{36}$$

As in Sect. 4 of this chapter, the corresponding homological operators will be denoted by $\mathcal{A} = L_A$ and by $\mathcal{B} = L_B$.

We assume that both A and B are normal matrices.

---

[5] In the general case, i.e. when A is not normal, the resonant terms will be those in $\mathrm{Ker}\,(L_{A^+})$.

The key, although trivial, observation is that the geometric relation $[X,Y] = 0$ does not depend on the coordinate system we are using. Therefore, if $\{f,s\} = 0$, we must also have $\{g,r\} = 0$.[6]

Another important, and again quite trivial, observation is that when we consider a Poincaré change of coordinates $x = y + h_k(y)$, the term F_k of order k in X changes according to $\mathcal{A}(h_k)$, but the term S_k of order k in Y changes according to the action of a different homological operator, i.e. according to $\mathcal{B}(h_k)$.

Thus, although when we choose $h_k + \ell_k$ [with $\ell_k \in \mathrm{Ker}\,(\mathcal{A})$] to generate the Poincaré transformation, we get the same transformation as that generated by h_k on the F_k, the transformation on S_k can be different. This means that the freedom left by the Poincaré prescription for construction of the normalizing transformation could, in principle, be used to take the symmetry vector field Y into some convenient form. This is indeed the case, as will be shown in this section.

Before stating precisely the result that can be obtained, we need to introduce the definition of the *Joint Normal Form*[7] (JNF), and recall some notions of linear algebra.

We recall that a real matrix T is semisimple if its complexification T_C can be diagonalized, and is normal if it commutes with its adjoint, i.e. if $[T,T^+] = 0$. For a normal matrix, $T : \mathrm{Ker}(T^+) \to \mathrm{Ker}(T^+)$. If T is normal we actually have $\mathrm{Ker}(T) = \mathrm{Ker}(T^+)$.

Any semisimple matrix can be transformed into a normal one by a linear transformation. If two semisimple matrices A, B commute, then they can be simultaneously diagonalized (by a linear, in general non-orthogonal, transformation), and so taken simultaneously to be normal. Thus, when considering such a pair of matrices, we can – with no loss of generality – assume them to be diagonal or, a fortiori, normal.

If we want to transform T into a real normal matrix, we just have to consider the transformation of T into a block diagonal matrix, the blocks corresponding to (complex conjugate) eigenvalues. It is easy to see that in this way we still get a (real) normal matrix.

Notice, however, that when the diagonalizing matrix M is not unitary, this transformation changes implicitly the scalar product, so that T semisimple will be normal with respect to the scalar product in the coördinates $y = Mx$. In the following, when we deal with semisimple matrices we will always assume, unless otherwise stated, that they have already been taken to be normal.

Definition 1. *Two vector fields $X, Y \in \mathcal{M}$, as in* (35),(36), *are in* Joint Normal Form *if A and B are semisimple, and both G_k and R_k are in* $\mathrm{Ker}\,(\mathcal{A}) \cap \mathrm{Ker}\,(\mathcal{B})$ *for all $k \geq 2$.*

---

[6] Clearly, when the bracket $\{.,.\}$ involves functions of the y, the gradient entering in its definition (9) is taken with respect to the y coordinates.

[7] We will follow mainly [95]; see also [172] for a different approach to a related problem.

We can then state our main result in this section as

Theorem 1. *Let the vector fields $X = f^i(x)\partial_i$ and $Y = s^i(x)\partial_i$ have a fixed point in $x_0 = 0$. Let they commute, $[X,Y] = 0$, and have normal semisimple linear parts $A = (Df)(0)$ and $B = (Ds)(0)$. Then, by means of a series of Poincaré transformations, they can be brought to Joint Normal Form.*

Proof. For our proof we follow [95], to which we refer the reader for further details. We use the power expansions (3), (11) for f and s, and the notation introduced so far. If X and Y commute, the equations (13) must be satisfied at any order. Let us assume that the theorem holds up to order $k-1$, i.e. F_n, S_n are in JNF up to order $k-1$.

This means in particular that in the symmetry equation

$$\mathcal{A}(S_k) - \mathcal{B}(F_k) = \psi_k \ ,$$

the ψ_k belongs to $\mathrm{Ker}(\mathcal{A}) \cap \mathrm{Ker}(\mathcal{B})$. Recall that here $\mathcal{A}, \mathcal{B}$ are the homological operators associated with A, B.

We then want to show that we can perform a Poincaré transformation taking them in JNF also up to order k. Let us first consider the problem of transforming F_k, and choose h_k as the solution of the corresponding homological equation. With this, F_k is changed into $\widehat{F}_k = F_k - \mathcal{A}(h_k)$, and S_k is also changed, into $\widehat{S}_k = S_k - \mathcal{B}(h_k)$. Obviously, ψ_k remains invariant. By definition, $\widehat{F}_k \in \mathrm{Ker}(\mathcal{A})$, and we can rewrite (24) in these new coordinates as

$$\mathcal{A}(\widehat{S}_k) - \mathcal{B}(\widehat{F}_k) = \sum_{j=2}^{k-1} \{\widehat{F}_j, \widehat{S}_{k-j+1}\} \equiv \psi_k \ . \tag{37}$$

Acting on this with $\mathcal{A}$ we get at once, as $[\mathcal{A}, \mathcal{B}] = 0$, that $\widehat{S}_k \in \mathrm{Ker}(\mathcal{A}^2)$. This also means, due to lemma 8, that $\widehat{S}_k \in \mathrm{Ker}(\mathcal{A})$, as we have assumed A to be normal.

Thus, $\mathcal{A}(\widehat{S}_k) = 0$, and the symmetry equation (since f is in NF up to order k) now reads just $\mathcal{B}(\widehat{F}_k) = -\psi_k$. We can act on this with $\mathcal{B}$, and get $\widehat{F}_k \in \mathrm{Ker}(\mathcal{B}^2)$. As we assumed B to be a normal matrix as well, we can just repeat the reasoning above, and conclude that $\mathcal{B}(\widehat{F}_k) = 0$. In this way we have shown that $\widehat{S}_k \in \mathrm{Ker}(\mathcal{A})$ and that $\widehat{F}_k \in \mathrm{Ker}(\mathcal{A}) \cap \mathrm{Ker}(\mathcal{B})$. The asymmetry between the two vector fields is just due to the fact that we choose h_k as the solution of the homological equation for f.

As already recalled above, the solution of this homological equation, i.e. h_k, is unique up to an element $\delta h_k \equiv \ell_k$ of $\mathrm{Ker}(\mathcal{A})$. Indeed, if we perform a new Poincaré transformation generated by $\ell_k \in \mathrm{Ker}(\mathcal{A})$, the $\widehat{F}_k$ remains unchanged. We write $\widetilde{F}_k = \widehat{F}_k$, and note explicitly that in this way we still have a $\widetilde{F}_k \in \mathrm{Ker}(\mathcal{A}) \cap \mathrm{Ker}(\mathcal{B})$, as we desire to prove.

However, in such a transformation the $\widehat{S}_k$ will not be invariant, but will instead be changed into $\widetilde{S}_k = \widehat{S}_k - \mathcal{B}(\ell_k)$. By suitably choosing ℓ_k, we can cancel out any term in the image of $\mathrm{Ker}(\mathcal{A})$ under $\mathcal{B}$. Using the invariant decomposition of

$\mathcal{V}_k$ introduced above we write $\text{Ker}(\mathcal{A}) = [\text{Ker}(\mathcal{A}) \cap \text{Ker}(\mathcal{B})] \cup [\text{Ker}(\mathcal{A}) \cap \text{Ran}(\mathcal{B})]$, and we conclude that $\mathcal{B}[\text{Ker}(\mathcal{A})]$ is just $\text{Ker}(\mathcal{A}) \cap \text{Ran}(\mathcal{B})$. Therefore, by suitably choosing ℓ_k we remain with $\widetilde{S}_k \in \text{Ker}(\mathcal{A}) \cap \text{Ker}(\mathcal{B})$.

We have thus proven that if X, Y are in JNF up to order $k-1$, we can find a Poincaré transformation of order k – which is just $h_k + \ell_k$ in the notation used here – such that they are also in JNF up to order k.

As under our assumptions X, Y are always in JNF up to order one (as they commute and have normal matrices for the linear parts), this concludes the proof. ⊙

Remark 1. In the above proof we have performed "simultaneously" the normalization of both f and s. It is important to remark that one could also proceed in a different way, that is putting, e.g., first f into NF: indeed, (37) for $k = 2$ gives immediately, as before, $\mathcal{A}(\widehat{S}_2) = 0$. On the other hand, an immediate application of the Jacobi identity shows that if $\widehat{F}_j,\ \widehat{S}_i \in \text{Ker}(\mathcal{A})$, the same is true for $\{\widehat{F}_j, \widehat{S}_i\}$. This allows us to proceed inductively again on (37) and to obtain $\widehat{F},\ \widehat{S} \in \text{Ker}(\mathcal{A})$, and so

$$\widehat{f}\ , \widehat{s} \in \text{Ker}(\mathcal{A})\ .$$

This implies in particular that – *once f is put into NF*, i.e. $f \to \widehat{f} = Ax + \widehat{F} \in \text{Ker}(\mathcal{A})$ – the symmetries of it are also symmetries of the linear part Ax. As noted in the proof, the series of transformations which normalize f are actually determined up to elements in $\text{Ker}(\mathcal{A})$, therefore, being $\widehat{S} \in \text{Ker}(\mathcal{A})$ and $\mathcal{B} : \text{Ker}(\mathcal{A}) \to \text{Ker}(\mathcal{A})$, one can still apply changes of coordinates to put also $\widehat{s}$ in NF with respect to $\mathcal{B}$ without changing the normality of $\widehat{f}$. The result $\widehat{f} \in \text{Ker}(\mathcal{B})$ automatically follows using the same argument, and the JNF is obtained as before. Clearly, the role of f and s can be reversed, and one can first perform a coordinate transformation which takes s into NF, $s \to \widehat{s} \in \text{Ker}(\mathcal{B})$. The same argument used above shows that this transformation will bring f into the form

$$f \to \widehat{f} \in \text{Ker}(\mathcal{B})$$

This fact will be useful in the problem of linearizing a DS, see Sect. 11 of this chapter.

In this theorem, we did not really use the interpretation of one of the vector fields as describing the dynamics of the system and the other describing a symmetry, rather we only used their commutation relation, i.e. the fact that they form an abelian two-dimensional algebra. From this point of view, it would be quite natural to also consider the case of non-commuting vector fields, i.e. of a general two-dimensional Lie algebra.

In the case of two vector fields, we can always reduce to the case $[X, Y] = cY$ by means of linear transformations. This in turn implies that their linear parts A and B satisfy $[A, B] = cB$. However, the assumption that A and B are semisimple does imply that in this case they actually commute.[8] For $c \neq 0$, this requires that

[8] This is easily seen, for example, by using the base in which B is diagonal.

$B = 0$. This is surely possible, but the JNF would then be just the usual normal form with respect to A alone. Thus, although it is possible to prove formally such a theorem for two-dimensional non-abelian algebras of vector fields having semisimple linear parts, this would be of no interest.

The generalization to arbitrary (non-abelian) algebras of vector fields is relevant and of interest when we have non-semisimple linear parts, or when we have semisimple linear parts and a Lie algebra of dimension $d \geq 3$ (as for the three-dimensional rotations).

In the next few sections, we will only consider the case of a single vector field Y being actually a symmetry of X, i.e. $[X, Y] = 0$, but in Sects. 9 and 10 of this chapter we will consider again the case of general algebras of vector fields.

Exercise 9. Prove that for $\mathcal{A}, \mathcal{B}$ commuting selfadjoint linear operators, one has $\mathcal{B}\,(\mathrm{Ker}(\mathcal{A})) = \mathrm{Ker}(\mathcal{A}) \cap \mathrm{Ran}(\mathcal{B})$.

Exercise 10. Use the Jacobi identity and the definition of $L_A \equiv \{Ax, .\}$ to show that for an algebra of matrices $\{A_1, ..., A_r\}$ satisfying $[A_i, A_j] = c_{ij}^k\ A_k$ the operators $L_i \equiv L_{A_i}$ satisfy $[L_i, L_j] = c_{ij}^k\ L_k$ with the same structure constants c_{ij}^k.

Exercise 11. Consider the following dynamical system X in $\mathbf{R}^3$, where $\rho^2 = x^2 + y^2$:

$$\begin{aligned} \dot{x} &= x - y\rho^2 \\ \dot{y} &= y + x\rho^2 \\ \dot{z} &= (y + x\rho^2)z \ . \end{aligned}$$

Determine the homological operator $\mathcal{A}$ associated with the linear part of such an X, and study its action on a generic vector function of $y = (x, y, z)$, written in components as $h(y) = (\alpha(y), \beta(y), \gamma(y))$. Prove that $h(y)$ is in $\mathrm{Ker}(\mathcal{A})$ if and only if

$$h(y) = \begin{pmatrix} a(z)x + b(z)y \\ c(z)x + d(z)y \\ e(z) \end{pmatrix}$$

where $a, ..., e$ are arbitrary polynomials in z. Check that this X admits the symmetry $Y = -y\partial_x + x\partial_y + xz\partial_z$, and provide a geometrical interpretation of the action of Y. Consider the homological operator $\mathcal{B}$ associated with the linear part of Y, and characterize the $h(y)$ which belong to $\mathrm{Ker}(\mathcal{B})$. Finally, use Theorem 1 to derive the possible JNF expansions for X and Y. (Hint: see [95].)

Exercise 12. For the same X, Y as in Exercise 11, perform the NF reduction up to order three.

Exercise 13. (See [96].) Consider the dynamical system X in $\mathbf{R}^3$ given (with $\rho^2 = x^2 + y^2$) by

$$\begin{aligned} \dot{x} &= -\,y + x\left(\rho^2 - ze^{-y}\right) \\ \dot{y} &= x + y\left(\rho^2 - ze^{-y}\right) \\ \dot{z} &= -\,z + \rho^2 e^y + yz\left(\rho^2 - ze^{-y}\right) + xz \ . \end{aligned}$$

Using the fact that this admits the symmetry $Y = y\partial_x - x\partial_y - xz\partial_z$, show that this can be reduced to a normal form which is symmetric under rotations around the z axis.

6. Linear Symmetries and Normal Forms

A special case of symmetries is given by *linear* symmetries, i.e. by the case where

$$Y = X_B = \left(B_{ij}x^j\right) \frac{\partial}{\partial x^i} \ . \tag{38}$$

It is this case that we briefly consider in this section. We have indeed

Corollary 1. *If the linear vector field $Y = X_B = B_{ij}x_j\partial_i$ is a symmetry for $X = f^i(x)\partial_i$, so that $\{Bx, f(x)\} = 0$, it is possible to normalize f by going to y coordinates so that $X = g^i(y)\partial_i$ and $\{By, g(y)\} = 0$. Then X_B is also a symmetry for g, and $g \in \mathrm{Ker}(\mathcal{A}) \cap \mathrm{Ker}(\mathcal{B})$.*

Proof. From $[A, B] = 0$ and $\{Bx, f\} = 0$ we have that $f \in \mathrm{Ker}(\mathcal{B})$ and that $\mathcal{A} : \mathrm{Ker}(\mathcal{B}) \to \mathrm{Ker}(\mathcal{B})$. This guarantees that we can normalize f – i.e. eliminate the terms which are not resonant with A, remaining in $\mathrm{Ker}(\mathcal{B})$. In this way, we get $g \in \mathrm{Ker}(\mathcal{A}) \cap \mathrm{Ker}(\mathcal{B})$ and $\{By, g(y)\} = 0$. ⊙

Corollary 2. *If $s(x) = Bx + S(x)$, with S the nonlinear part of s, and $Y = s^i(x)\partial_i$ is a symmetry of $X = f^i(x)\partial_i$, when the vector fields X, Y are put in Joint Normal Form we have that X_B is a linear symmetry of X.*

Proof. If X, Y are in JNF, we have in particular that $f \in \mathrm{Ker}(\mathcal{B})$. ⊙

When we perform the Poincaré transformations needed to transform X into its normal form, i.e. when we go to the y coordinates in the notation introduced above, there seems to be no reason, a priori, why the Y should keep its linear form. In general, if we go from coordinates x to coordinates y with $x = \varphi(y)$, and with Λ the matrix $\Lambda_{ij} = (\partial y^i / \partial x^j)$, we have

$$Y = \left[\Lambda_{ij} B_{jk} \varphi^k(y)\right] \frac{\partial}{\partial y^k} \ .$$

However, recall that we have some freedom in the choice of the Poincaré transformations, i.e. of the h_k generating them. It is actually possible to prove the following result that we quote from [190]:

Theorem 2. *If $X \in \mathcal{M}$ commutes with a linear vector field $Y = X_B = (B_{ij}x^j)\partial_i \in \mathcal{M}$, then it is possible to find a normalizing series of Poincaré transformations with generators $h_k \in \mathrm{Ker}(\mathcal{B})$, so that in the new coordinates y, X is taken into normal form and Y is left unchanged, i.e. $Y = (B_{ij}y^j)\partial_i$.*

The proof of this relies on a similar result by Ruelle [278] dealing with the center manifold mapping, and is given e.g. in [121,190]; see also [38,39,40]. See also Exercise 15 below.

Let us note that Theorem 2 is *not* a special case of Theorem 1 and of the above Corollaries. Indeed by that result, if B is resonant, we are by no means guaranteed that we can normalize X and still leave Y unchanged. On the other hand, if B does not admit any resonance, we know that the normal form of Y is just its linear form, and Theorem 1 does provide the same result.

Note also that if we interpret the normalizing transformation as the time one action of a vector field Φ [47] (see also Chaps. II and V), Theorem 2 shows that Φ can be chosen to admit Y as a symmetry; see the exercises below.

Exercise 14. A vector field $X = f^i(x)\partial_i : M \to \mathrm{T}M$ is said to be *reversible* if there is a map $R : M \to M$ such that $R^2 = I$, and such that $f(Rx) = (DR)f(x)$. Consider the case where R is a matrix so that $f^i(Rx) = R^i{}_j f^j(x)$, and show that when we put X in normal form, and write this as $X = \widetilde{f}^i(x)\partial_i$, we still have $\widetilde{f}^i(Rx) = R^i{}_j \widetilde{f}^j(x)$. Can we say anything for the case where R is a nonlinear map ?

Exercise 15. Show that if $X = f^i(x)\partial_x$ commutes with a linear vector field $Y = B^i{}_j x^j \partial_i$, the generators h_k of the Poincaré transformations needed to take X into NF can be taken to be symmetric under Y. (Hint: The homological equations to solve are of the form $\mathcal{A}(h_k) = G_k \equiv \pi F_k$, with $G_k \in \mathrm{Ker}(\mathcal{A}) \cap \mathrm{Ker}(\mathcal{B})$.)

7. On Linear and Nonlinear Symmetries

The relation between linear and nonlinear symmetries, especially the "quasi-linear" symmetries to be considered in a moment, are sometimes confusing; in this section we would like to clarify some points in this respect.

First of all, we denote by $\mathcal{L}_X \subseteq \mathcal{G}_X$ the set of linear symmetries of X (and we use the shorthand notation $\mathcal{L} \equiv \mathcal{L}_X$), i.e.

$$\mathcal{L} \;=\; \{Y = s^i(x)\partial_i \in \mathcal{G} \;:\; s^i(x) = B_{ij}x^j\} \;. \tag{39}$$

We also recall that by $\mathcal{I} \equiv \mathcal{I}_X$ we denote the (analytic) constants of motion (or first integrals) of X, i.e. the functions $\rho : M \to \mathbf{R}$ such that $X(\rho) = 0$.

We will then denote by $\mathcal{Q}_X \subseteq \mathcal{G}_X$ (and use again $\mathcal{Q} \equiv \mathcal{Q}_X$ for ease of notation) the set of quasi-linear symmetries of X [204]. By this we mean linear combinations of terms $Y \in \mathcal{G}$ of the form

$$Y = s^i(x)\partial_i \quad \text{with} \quad s^i(x) = \mu(x)\; B_{ij}x^j \;,\; \mu(x) \in \mathcal{I} \;,\; B \in \mathcal{L} \;. \tag{40}$$

Clearly, $\mathcal{Q} = \mathcal{I} \otimes \mathcal{L} \subseteq \mathcal{G}$. Notice that $\mathcal{Q}$ is a module over the algebra $\mathcal{I}$ (and the notation $\mathcal{I} \otimes \mathcal{L}$ means, as mentioned above, linear combinations of elements of $\mathcal{L}$ with coefficients in $\mathcal{I}$). We would like to point out that in general $\mathcal{Q} \neq \mathcal{G}$, i.e. not all the $Y \in \mathcal{G}$ are of the quasi-linear form.

It is worth considering in further detail the relation between $\mathcal{G}$ and the $\mathcal{L}$, $\mathcal{Q}$ defined above.

It is also useful to consider $\mathcal{G}_Z, \mathcal{L}_Z$, etc. for linear vector fields Z, and in particular for the vector field $A_{ij}x^j\partial_i \equiv (Ax)\cdot\nabla$ corresponding to the linearization of

X around its fixed point $x_0 = 0$. In this case we will simply write $\mathcal{G}_A, \mathcal{L}_A, \mathcal{Q}_A, \mathcal{I}_A$ (similarly to what we do for vector fields).

We will also find it useful to separate the linear and nonlinear terms in the power series expansion of our functions; thus, we write

$$f(x) = Ax + F(x) \quad , \quad s(x) = Bx + S(x) \ . \tag{41}$$

By writing down series expansions and looking at lowest order terms, we obtain easily the following:[9]

Lemma 9. *If $[X,Y] = 0$ and A, B are the linearizations of X, Y around their common fixed point $x_0 = 0$, then $[A, B] = 0$.*

By applying this to the case $Y = (B_{ij}x^j)\partial_i$, i.e. $Y \in \mathcal{L}_X$, we obtain

Lemma 10. *If $A = (Df)(0)$ is the linearization of X at its fixed point $x_0 = 0$, then $\mathcal{L}_X \subseteq \mathcal{L}_A$.*

One should avoid confusions between linear symmetries of the dynamical system and symmetries of its linearization, which do not extend in general to symmetries of the full system.

It should be emphasized that, when it happens that $\mathcal{G}_X = \mathcal{Q}_X$, or $\mathcal{I}_X = \mathcal{I}_A$, or $\mathcal{G}_X = \mathcal{G}_A$, then the analysis of symmetry and symmetry-related properties of X becomes easier, and stronger results hold. Let us briefly point out two (relevant) cases.

If A is non-resonant, then X can be reduced to its linear part A (which corresponds to the normal form). In this case, as the algebra $\mathcal{G}_X$ is the same before and after the normalizing transformation (although with different coordinate representations), it follows that $\mathcal{G}_X = \mathcal{G}_A$.

Consider the case where $X = f^i(x)\partial_i$ can be transformed by Poincaré transformations into the form (which is not necessarily its normal form) $X = q^i(y)\partial_i$, where

$$q(x) = (1 + \mu(y)) \ Ay \tag{42}$$

with $\mu(y)$ a real-valued scalar function (see also Chap. VI for the relevance of this case). We say in this case that X is in *quasi-linear* form. It is clear that (42) implies $\mathcal{I}_X = \mathcal{I}_A$; it is also easy to be convinced that $\mathcal{G}_X \neq \mathcal{G}_A$, as the identity matrix belongs to $\mathcal{L}_A \subseteq \mathcal{G}_A$ but does not belong (unless $\mu(y) \equiv 0$) to $\mathcal{L}_X \subseteq \mathcal{G}_X$.

---

[9] This was actually already used above, in the proof of theorem 1.

8. Symmetry for Systems in Normal Form

No definite relation exists, in general, between $\mathcal{G}_X$ and $\mathcal{G}_A$ or between $\mathcal{I}_X$ and $\mathcal{I}_A$; but if X is in normal form, one has some interesting and useful results [96,330]:

Lemma 11. *If X is in normal form, any constant of motion of X must also be a constant of motion of its linearization A, i.e. $\mathcal{I}_X \subseteq \mathcal{I}_A$.*

However, in general $\mathcal{I}_X \neq \mathcal{I}_A$, even if X is in normal form. Also, if $[B, A] = 0$ in general $\mu(x)Bx$ does not belong to $\mathcal{G}$, even for $\mu \in \mathcal{I}_X$ (unless $X_B \in \mathcal{L}_X$ as well), nor to $\mathcal{G}_A$ (unless $\mu \in \mathcal{I}_A$); see however below in this section.

Lemma 12. *If X is in NF, then any $Y \in \mathcal{G}_X$ is also a symmetry for its linear part A, i.e. $\mathcal{G}_X \subseteq \mathcal{G}_A$.*

This result can be obtained using the proof of Theorem 1, and can be a considerable help in the symmetry analysis, in that it allows restricting our search for $Y \in \mathcal{G}_X$ to $\mathcal{G}_A$ rather than considering the full $\mathcal{M}$.

Similarly, Lemmata 10-11 can be useful in the determination of the sets $\mathcal{L}_X, \mathcal{I}_X$, in that we can first solve the easier problem of determining $\mathcal{L}_A, \mathcal{I}_A$, and then look for $\mathcal{L}_X, \mathcal{I}_X$ in the class of vector fields $\mathcal{L}_A$ and of the functions in $\mathcal{I}_A$, rather than in the full set of linear vector fields on M and, respectively, in the full $\mathcal{F}$.

Moreover, $\mathcal{G}_A$ can be determined in a relatively simple way, by solving the system of quasi-linear non-homogeneous first order PDEs $\{Ax, g\} = 0$, which are written explicitly as

$$\left(A_{ij}\, x^j\right) \frac{\partial g^k}{\partial x^i} = A_{kj}\, g^j \ . \tag{43}$$

By considering this, and introducing the set $\mathcal{I}_A^*$ of the meromorphic (i.e., fractions of formal power series) constants of motion of the linear problem $\dot{x} = Ax$, one can obtain [96,121,330] the following result:

Lemma 13. *$\mathcal{G}_A$ is the set of all formal power series in $\mathcal{I}_A^* \otimes \mathcal{L}_A$. In a more explicit form, since $\mathcal{G}_A \equiv \mathrm{Ker}(\mathcal{A})$, one has that the resonant terms, $F(x) \in \mathrm{Ker}(\mathcal{A})$, are power series of the form $F(x) = K\big(\rho(x)\big) \cdot x$, where K is a matrix commuting with A when K is written in terms of its real entries K_{ij}, and where $K\big(\rho(x)\big)$ is the same matrix in which the entries K_{ij} are replaced by functions of the constants of motion $\rho = \rho(x) \in \mathcal{I}_A^*$. The set of the VF in $\mathcal{G}_A$ is naturally a Lie algebra.*

The matrices $K(\rho(x))$ mentioned in this Lemma (and in Lemma 17 below, Sect. 10 of this chapter) can be characterized more precisely as follows: If the set $\mathcal{L}_A$ is a Lie algebra (thus, a real linear space) spanned by the matrices $\{M_1, ..., M_r\}$, then $K(\rho(x))$ will belong to a module over $\mathcal{I}_A^*$ generated by the same matrices $\{M_1, ..., M_r\}$.

We summarize our discussion for dynamical systems in normal form in the following proposition:

Proposition. *Let $X = f^i(x)\partial_i$ be a vector field in normal form, and let $A = (Df)(0)$ be the normal matrix corresponding to its linear part. Then, $\mathcal{L}_X \subseteq \mathcal{G}_X \subseteq \mathcal{G}_A$; and $\mathcal{L}_X \subseteq \mathcal{L}_A \subseteq \mathcal{G}_A$.*

It should be mentioned that Kodama has recently considered the problem of determining $\mathcal{G}_A$ from a more algebraic standpoint [204]. In the same work, Kodama also observed that $\mathcal{G}_A$, considered as an algebra, is not only infinite-dimensional, but has the natural structure of a graded Virasoro algebra.

We would like to emphasize once again that the above results were given for X in normal form. They can obviously be no longer true if X is not in NF, as the following example shows: Consider the two-dimensional system, which is *not* in normal form,

$$\begin{aligned} \dot{x}_1 &= -\, x_1 x_2 \\ \dot{x}_2 &= \; x_2\ . \end{aligned} \tag{44}$$

It is easily seen that $\mu = x_1 \exp(x_2)$ is a constant of motion of the DS, but not of the linearized problem, i.e. $\mu \in \mathcal{I}_X$, but $\mu \notin \mathcal{I}_A$. Similarly, we have that $Y = x_1^2 \exp(x_2)(\partial/\partial x_1) \in \mathcal{G}_X$ but $Y \notin \mathcal{G}_A$.

Exercise 16. Consider the two-dimensional dynamical system

$$\dot{x} = \alpha(\rho) J x \equiv f(x) \quad , \quad J = \begin{pmatrix} 0 & -1 \\ 1 & 0 \end{pmatrix} \ ; \ \rho = x_1^2 + x_2^2$$

with $\alpha(0) \neq 0$ and $\alpha(\rho) \neq$ const. Show that in this case any $B \in \mathcal{L}_X$ is of the form $B = cJ$ with c a real constant, and that any $Y \in \mathcal{G}_X$ can be written in the form $Y = \mu(\rho)\ J_{ik} x^k \partial_i$ with μ a smooth function. Show that $\mathcal{I}_A = \mathcal{I}_X$ and that $\mathcal{G}_X = \mathcal{I}_X \otimes \mathcal{L}_X = \mathcal{Q}_X$. Show also that $\mathcal{L}_X \neq \mathcal{L}_A$ by pointing out a matrix which belongs to $\mathcal{L}_A$ but not to $\mathcal{L}_X$, and finally that, in this case, $\mathcal{G}_A = \mathcal{I}_A \otimes \mathcal{L}_A = \mathcal{Q}_A$, but $\mathcal{G}_A \neq \mathcal{G}_X$. Consider the following modification of this example:

$$\dot{x} = \alpha(\rho) J x + \beta(\rho) x \equiv f(x)$$

(notice that X is again in NF). Show that $\mathcal{I}_X \neq \mathcal{I}_A$ by pointing out a scalar function $\tau : \mathbf{R}^2 \to \mathbf{R}$ which is a constant of motion of the linear part $\dot{x} = Jx$, but not of the full problem. Show that, actually, both $\mathcal{I}_X$ and $\mathcal{I}_X^*$ are trivial.

Exercise 17. Consider the two-dimensional linear problem $\dot{x} = Ax$ where $A = \mathrm{diag}(1,2)$. Show that $\mathcal{I}_A$ is trivial, but $\mathcal{I}_A^*$ is not. Show also that the nonlinear terms in $\mathcal{G}_A = \mathrm{Ker}(\mathcal{A})$ are the multiples of just one vector field, and that $\mathcal{G}_A \neq \mathcal{I}_A \otimes \mathcal{L}_A = \mathcal{L}_A$.

9. Reduction to Normal Form of a Nilpotent Lie Algebra

Up to now, we have considered a vector field X defining a dynamical system, and *one* symmetry vector field Y. In this section, we start discussing the case of a symmetry algebra, i.e. of several symmetry vector fields $Y_1, Y_2, ...Y_r$.

It is actually convenient to drop the distinction between the vector field defining the DS and the symmetry vector fields. Thus, we simply consider an algebra $\mathcal{G}$ of vector fields $X_i \in \mathcal{M}$. If this is the symmetry algebra of one of the vector fields, say X_0, we have that $[X_0, X_i] = 0$ for all the i, i.e. that X_0 belongs to the center of the algebra, $X_0 \in \mathcal{C}(\mathcal{G})$.

First of all, we consider the case of an *abelian* Lie algebra of vector fields.[10] It will turn out that the nilpotent case is actually equivalent to this.

Theorem 3. *Let $\{X_1, \ldots, X_r\}$ span an abelian Lie algebra $\mathcal{G}$ under $[.,.]$, and assume that the matrices $A_{(i)}$ identifying the linear parts of the vector fields $X_j = f^i_{(j)}(x)\partial_i$, $(Df_{(j)})(0) = A_{(j)}$ are normal. Then by a series of Poincaré transformations, $\{X_1, \ldots, X_r\}$ can be put in Joint Normal Form; i.e.*

$$f_{(j)} \to \widetilde{f}_{(j)} \in \mathcal{K} = \bigcap_{i=0}^{k} \mathrm{Ker}(\mathcal{A}_i) \qquad \forall j = 1, \ldots, k$$

where the $\mathcal{A}_i$ are the homological operators associated with the matrices $A_{(i)}$.

Proof: The proof is quite similar to the one for Theorem 1, and is sketched in [96]. We will write each of the $f_{(i)}$ appearing in the X_i as a power expansion in homogeneous terms, i.e. $f_{(i)}(x) = \sum_{k=1}^{\infty} F_k^{(i)}(x)$, where $F_1^{(i)} = A_{(i)}$. The homological operator associated with $A_{(i)}$ will be denoted as $\mathcal{A}_i$.

We also denote by $\psi_k^{(i,j)}$ the analog of the ψ_k considered before, and built with the vectors X_i and X_j. That is,

$$\psi_k^{(i,j)} = - \sum_{m=2}^{k-2} \{F_m^{(i)} \, , \, F_{k-m}^{(j)}\} \, . \tag{45}$$

The condition $[X_i, X_j] = 0$ reads then at order k:

$$\mathcal{A}_i \left(F_k^{(j)}\right) - \mathcal{A}_j \left(F_k^{(i)}\right) - \psi_k^{(i,j)} = 0 \, . \tag{46}$$

Assume that $\mathcal{G}$ is in JNF up to order $k-1$. Let us perform a Poincaré transformation with generator $h_k^{(1)}$ chosen to have X_1 in NF up to order k, i.e. so that $\widehat{F}_k^{(1)} = F_k^{(1)} - \mathcal{A}_1(h_k^{(1)}) \in \mathrm{Ker}(\mathcal{A}_1)$. With this transformation, the other $F_k^{(i)}$ are transformed into $\widehat{F}_k^{(i)} = F_k^{(i)} - \mathcal{A}_i(h_k^{(1)})$, while the $\psi_k^{(i,j)}$ are unchanged.

---

[10]Here we assume normality of the relevant matrices, but it should be recalled that we could also assume semisimplicity, as commuting semisimple matrices can be simultaneously diagonalized.

Acting with $\mathcal{A}_1$ on the equations $\mathcal{A}_i(\widehat{F}_k^{(1)}) - \mathcal{A}_1(\widehat{F}_k^{(i)}) = \psi_k^{(i,1)}$, using again $[\mathcal{A}_i, \mathcal{A}_j] = 0$, and the normality of $A_{(1)}$ as in the proof of Theorem 1, we get that $\widehat{F}_k^{(i)} \in \mathrm{Ker}(\mathcal{A}_1)$. This also means that we actually have $\mathcal{A}_i(\widehat{F}_k^{(1)}) = \psi_k^{(i,1)}$. By acting on this with $\mathcal{A}_i$, using the normality of $A_{(i)}$, and repeating this for all the i, we also conclude that $\widehat{F}_k^{(1)} \in \mathrm{Ker}(\mathcal{A}_1) \cap ... \cap \mathrm{Ker}(\mathcal{A}_r)$.

We can now perform a second Poincaré transformation with generator $h_k^{(2)} \in \mathrm{Ker}(\mathcal{A}_1)$, which will leave $\widehat{F}_k^{(1)}$ unchanged. Proceeding as in the proof of Theorem 1, we observe that by suitably choosing $h_k^{(2)}$ we can eliminate any term of $\widehat{F}_k^{(2)}$ which lies in the image of $\mathrm{Ker}(\mathcal{A}_1)$ under $\mathcal{A}_2$, i.e. in $\mathrm{Ker}(\mathcal{A}_1) \cap \mathrm{Ran}(\mathcal{A}_2)$.

From now on we will denote by $\widehat{F}_k^{(i)}$ the terms having undergone the second Poincaré transformation, i.e. $\widehat{F}_k^{(i)} \to \widehat{F}_k^{(i)} - \mathcal{A}_i(h_k^{(2)})$.

Thus, after this second transformation we will have a $\widehat{F}_k^{(2)} \in \mathrm{Ker}(\mathcal{A}_1) \cap \mathrm{Ker}(\mathcal{A}_2)$. Let us now act with $\mathcal{A}_2$ on the equations $\mathcal{A}_i(\widehat{F}_k^{(2)}) - \mathcal{A}_2(\widehat{F}_k^{(i)}) = \psi_k^{(i,2)}$. We get, exactly in the same way as before, that $\widehat{F}_k^{(i)} \in \mathrm{Ker}(\mathcal{A}_2)$, so that indeed $\mathcal{A}_i(\widehat{F}_k^{(2)}) = \psi_k^{(i,2)}$ and, by applying again $\mathcal{A}_i$ we get, due to the normality of $A_{(i)}$, that $\widehat{F}^{(2)} \in \mathrm{Ker}(\mathcal{A}_i)$ for all the $\mathcal{A}_i$.

We can repeat the same procedure for new Poincaré transformations $h_k^{(3)}, ...$ up to $h_k^{(r)}$, where $h_k^{(j)} \in \bigcap_{i=1}^{j-1} \mathrm{Ker}(\mathcal{A}_i)$ is chosen to take $\widehat{F}_k^{(j)}$ into NF. At the end of such a procedure the initial $F_k^{(i)}$ are transformed into $\widetilde{F}_k^{(i)}$ which belong, as required, to $\mathcal{K} = \bigcap_{i=1}^{r} \mathrm{Ker}(\mathcal{A}_i)$.

As, under our assumptions, the $\{X_1, ..., X_r\}$ are always in JNF at order $k = 1$, this concludes the proof for the abelian case. ⊙

It is very hard, if not impossible, to obtain results for generic Lie algebras, and indeed it is quite clear that one cannot expect results as general as in the abelian case [9, 96].[11] A significant exception to this will be the case of *nilpotent* algebras (see also [107] for a group-theoretical approach), in which we can recover essentially the same results obtained in the abelian case.[12]

However, an extension of Theorem 3 to the nilpotent case should be considered with some care, as the only nilpotent algebras of semisimple matrices are abelian. On the other hand, similarly to what has already been remarked after obtaining Theorem 1, we could have a non-abelian nilpotent algebra of vector fields with linear parts given by semisimple matrices, provided that some of these vanish.

Indeed, although $[X_i, X_j] = c_{ij}^k X_k$ necessarily implies that $[A_i, A_j] = c_{ij}^k A_k$, it could happen that all the A_k for which there exists a nonzero c_{ij}^k do vanish.

---

[11] It should be pointed out that the theorem given in [96] should actually require, in the case of non-nilpotent algebras, a more careful consideration of the various possible cases. See Sect.12 of this chapter, where we will show that, in fact, some "partial" JNF can be obtained in particular cases.

[12] See [181] for the case of a semisimple Lie algebra.

Although in this case $\text{Ker}(\mathcal{A}_k)$ would be just the whole space, we do still need to consider the corresponding X_k in order to have a closed Lie algebra.

With this remark, and in view of the fact that the proof of Theorem 3 is based on properties of the algebra of the A_i's and of the corresponding homological operators, it is to be expected that the proof for the nilpotent case will not differ from that for the abelian case. This is indeed what happens, and one has the following result [96]:

Theorem 4. *Let the vector fields $\{X_1, \ldots, X_r\}$ form a nilpotent Lie algebra $\mathcal{G}$ under $[.,.]$, and assume that the matrices $A_{(i)}$ identifying the linear parts of the vector fields $X_j = f^i_{(j)}(x)\partial_i$, $(Df_{(j)})(0) = A_{(j)}$ are normal. Denote by $\mathcal{A}_i$ the homological operators associated with the matrices $A_{(i)}$. Then, by a series of Poincaré transformations, $\{X_1, \ldots, X_r\}$ can be put in Joint Normal Form; i.e.*

$$f_{(j)} \to \widetilde{f}_{(j)} \in \mathcal{K} = \bigcap_{i=0}^{k} \text{Ker}(\mathcal{A}_i) \qquad \forall j = 1, \ldots, k \, .$$

Proof: Again, the proof is quite similar to that for Theorem 1, and is sketched in [96]. We use the same notation as in the proof of Theorem 3.

In this case, the condition $[X_i, X_j] = c^k_{ij} X_k$ (with c^k_{ij} the structure constants of the Lie algebra) reads at order k:

$$\mathcal{A}_i \left(F_k^{(j)}\right) - \mathcal{A}_j \left(F_k^{(i)}\right) - \psi_k^{(i,j)} = c^\ell_{ij} F_k^{(\ell)} \, . \tag{46'}$$

The abelian case has been proved as Theorem 3. The extension to the nilpotent case follows at once by considering iterated commutation relations: If the rank of nilpotency of the algebra is p, we have that $F_k^{(i)} \in \text{Ker}(\mathcal{A}_j^p)$ implies, due to the normality of the $A_{(i)}$, that actually $F_k^{(i)} \in \text{Ker}(\mathcal{A}_j)$. We can then repeat exactly the same reasoning and complete the proof. ⊙

It should be mentioned that the linear parts of some of the vector fields could vanish. This is e.g. the case for the algebra spanned by $X = -x^2 d/dx$ and $Y = x d/dx$.

Corollary 3. *If the algebra $\mathcal{G}$ of vector fields $\{X_1, ..., X_n\}$ contains a nilpotent subalgebra $\mathcal{G}^*$, then the series of the f_i corresponding to the vector fields X_i spanning this $\mathcal{G}^*$ can be put in Joint Normal Form.*

Proof. This is proven by considering transformations chosen so as to normalize the appropriate subalgebra of $\mathcal{G}$. ⊙

Exercise 18. Let Φ be an invertible diffeomorphism of $M \subseteq \mathbf{R}^n$; consider the vector fields $X_{(k)} = f^i_{(k)}(x)\partial_i$ (for $k = 0, 1, 2, ...$) given by $f_{(k)}(x) = \left[(D\Phi)^{-1} A^k \Phi\right](x)$, where A is a normal matrix (and A^k is its k-th power). Do these matrices form an algebra ? What can we say about their transformation to joint normal form ?

10. Non-semisimple Normal Forms

In our discussion, we have so far relied on the assumption that the relevant matrices $A_{(i)}$, identifying the linear parts of the vector fields X_i we consider, are normal, and we can deal with the homological operators $\mathcal{A} = L_A$ rather than with L_{A^+}. Also, we have at several stages used the fact that $\mathrm{Ker}(\mathcal{A}) \cap \mathrm{Ran}(\mathcal{A}) = \{0\}$; this again was guaranteed by the assumption on the normality of A.

In this section, we consider how the main results obtained so far are modified for the non-normal case. The reader is also referred to [12,16,121,190,330,331] for details on normal forms with non-normal (non-semisimple) linear parts. We find that, actually, most of our results continue to hold, albeit in a weaker form (see also [345]).

A classical result in linear algebra (see e.g. [183] or [250]) is that

Lemma 14 (S+N decomposition). *A generic matrix A is uniquely decomposed as*

$$A = A_s + A_n \tag{47}$$

where A_s is semisimple (i.e. diagonalizable), A_n is nilpotent, and these commute, $[A_s, A_n] = 0$.

We will call the unique decomposition described in this lemma the *S+N decomposition*. A stronger result [9,107] holds for the case of a set of matrices spanning a nilpotent Lie algebra:[13]

Lemma 15. *Let the matrices $A^{(i)}$, $i = 1, ..., n$, span a nilpotent Lie algebra. Then there are semisimple matrices $A_s^{(i)}$ and nilpotent matrices $A_n^{(i)}$, $i = 1, ..., n$, such that $A^{(i)} = A_s^{(i)} + A_n^{(i)}$, and $\left[A_s^{(i)}, A_n^{(j)}\right] = 0 \ \forall i, j = 1, ..., n$.*

For the case of two commuting matrices, we have the following lemma (which could also be obtained as a corollary to lemma 15):

Lemma 16. *If the matrices A and B commute, $[A, B] = 0$, and have S+N decompositions $A = A_s + A_n$ and $B = B_s + B_n$, then $[A_s, B] = [A, B_s] = 0$.*

Proof. Write either one of the matrices A, B in Jordan form and impose commutation in generalized eigenspaces. ⊙

We will also often assume for convenience that the semisimple part A_s is a normal matrix, $[A, A^+] = 0$; this indeed can be obtained by means of a linear transformation (e.g. the one diagonalizing A_s). If we have matrices A, B with commuting semisimple parts (in particular, but not only, if $[A, B] = 0$), then A_s and B_s can be simultaneously diagonalized, and thus we can in full generality assume that A_s and B_s are both normal. The same holds for an algebra of matrices whose semisimple parts span an abelian algebra.

---

[13] This can also be proved directly by writing the matrices of $\mathcal{G}_n$ (see the Appendix to this chapter) in canonical form, and then working it back to the full $\mathcal{G}$.

In the case of matrices with a nilpotent part, the definition of resonant terms will be slightly modified. First of all, it should be stressed that, in general (i.e. if we take a general scalar product), the homological operator associated with the adjoint A^+ of the matrix A would *not* be the adjoint $\mathcal{A}^+$ of the homological operator $\mathcal{A}$ associated with A. When we adopt in the space $\mathcal{V}$ of vector power series the standard scalar product [121] introduced in Sect. 5 of Chap. II, these are the same [121,190]), so that the notation $\mathcal{A}^+ = \{A^+x, .\}$ is justified. Considering now the homological operators

$$\mathcal{A}_s = \{A_s x, \cdot\} \quad , \quad \mathcal{A}^+ = \{A^+ x, \cdot\} \; , \tag{48}$$

we say that a vector function $f(x) \in \mathcal{V}$ is *resonant* with the matrix A if and only if $f(x) \in \mathrm{Ker}(\mathcal{A}^+)$ or, equivalently, if and only if $\{A^+x, f(x)\} = 0$.

For the set $\mathrm{Ker}(\mathcal{A}^+)$ of resonant vector functions, we have the following result, similar to Lemma 13:

Lemma 17. *The terms $F(x) \in \mathrm{Ker}(\mathcal{A}^+)$ are power series of the form $F(x) = K(\rho(x)) \cdot x$, where K is a matrix which commutes with A^+ written in terms of the real constants K_{ij}, and $K(\rho(x))$ is the same matrix where the K_{ij} are replaced by functions of the constants of motion $\rho(x) \in \mathcal{I}^*_{A^+}$ for the linear problem $\dot{x} = A^+x$.*

Explicit computations of the time-independent constants of motion of linear problems of the type $\dot{x} = Ax$ with non-semisimple A, easily show, using also Lemma 17 and the analyticity of the elements in $\mathrm{Ker}(\mathcal{A})$ and $\mathrm{Ker}(\mathcal{A}^+)$, the following lemma:[14]

Lemma 18. *With the above notations,* $\mathrm{Ker}(\mathcal{A}) \subseteq \mathrm{Ker}(\mathcal{A}_s)$ *and* $\mathrm{Ker}(\mathcal{A}^+) \subseteq \mathrm{Ker}(\mathcal{A}_s)$*; if moreover A_s is normal, then* $\mathrm{Ker}(\mathcal{A}_s) = \mathrm{Ker}(\mathcal{A}_s^+)$.

Remark 2. It is clear that the Poincaré-Dulac theorem, reformulated in the non-semisimple case, states that, given any VF $f = Ax + F$, the nonlinear part can be transformed by means of a formal Poincaré transformation into NF with respect to $\mathcal{A}^+$, i.e. $F \to \widehat{F} \in \mathrm{Ker}(\mathcal{A}^+)$. Notice that, in this case, the resonance condition $\mathcal{A}^+(\widehat{F}) = 0$ cannot be extended to the linear term: $Ax \notin \mathrm{Ker}(\mathcal{A}^+)$. However, $Ax \in \mathrm{Ker}(\mathcal{A}_s)$, indeed $[A, A_s] = 0$ thanks to Lemma 18, therefore one can have $\widehat{f} \in \mathrm{Ker}(\mathcal{A}_s)$.

With these, we define the Semisimple Joint Normal Form (SJNF) and X-Joint Normal Form (X-JNF) for linear parts given by general (i.e. non-semisimple) matrices as follows:

Definition 2. *Let us consider two vector fields $X = f^i(x)\partial_i$ and $Y = g^i(x)\partial_i$, and let $A = (Df)(0)$, $B = (Dg)(0)$ be their linear parts. Let these have the S+N decompositions $A = A_s + A_n$, $B = B_s + B_n$. Let the nonlinear parts of f and g be respectively F and G, i.e. $f(x) = Ax + F(x)$, $g(x) = Bx + G(x)$. We say that*

[14] This can also be obtained directly from algebraic properties of A, A^+, A_s and is well known [250] in the context of the theory of replicas [12,250].

X, Y are in Semisimple Joint Normal Form *if and only if both F and G are in* $\mathrm{Ker}(\mathcal{A}_s) \cap \mathrm{Ker}(\mathcal{B}_s)$.

Notice that when A, B are semisimple (and normal) matrices, this definition coincides with Definition 1.

Definition 3. *Let us consider two vector fields $X = f^i(x)\partial_i$ and $Y = g^i(x)\partial_i$, and let $A = (Df)(0)$, $B = (Dg)(0)$ be their linear parts. Let these have the S+N decompositions $A = A_s + A_n$, $B = B_s + B_n$. Let the nonlinear parts of f and g be respectively F and G, i.e. $f(x) = Ax + F(x)$, $g(x) = Bx + G(x)$. We say that X, Y are in X-*Joint Normal Form *(X-JNF) if and only if*

$$F \in \mathrm{Ker}(\mathcal{A}^+) \cap \mathrm{Ker}(\mathcal{B}_s) \qquad \text{and} \qquad g \in \mathrm{Ker}(\mathcal{A}_s) \cap \mathrm{Ker}(\mathcal{B}_s)$$

or clearly in Y-JNF if the role of f and g is reversed.

With these preliminary results, one can provide counterparts to the results obtained above under the normality assumption, for the case of non-normal linear parts. Theorem 6 below will give the most general result, but we believe that discussing first the special case of two vector fields, rather than that of a general r-dimensional algebra, will help in getting an understanding of the situation. As already remarked, in the case of two vector fields we can always reduce to $[X, Y] = cX$. If $c = 0$ the algebra is abelian, and therefore nilpotent, while for $c \neq 0$ the algebra is solvable but not nilpotent.

Theorem 5. *Let the vector fields $X = f^i(x)\partial_i$ and $Y = g^i(x)\partial_i$ have a common fixed point in $x_0 = 0$, and satisfy $[X, Y] = 0$. Let their linear parts $A = (Df)(x_0)$ and $B = (Dg)(x_0)$ have the S+N decompositions $A = A_s + A_n$, $B = B_s + B_n$, where the semisimple parts A_s and B_s are also normal. Then they can be brought to Semisimple Joint Normal Form by means of a formal series of Poincaré transformations. It is also possible, by means of a formal series of Poincaré transformations, to take them to X-Joint or Y-Joint Normal Form.*

Proof. The idea of the proof is similar to that for Theorem 1. Indeed, from the commutation property $[X, Y] = 0$ one obtains, exactly as in (37),

$$\mathcal{A}(G_k) - \mathcal{B}(F_k) = \psi_k$$

where ψ_k is a combination of terms G_h, F_m with $h, m < k$ as in (37). Let now $G_k \in \mathrm{Ker}(\mathcal{B}_s)$. By applying the operator $\mathcal{B}_s$ to this equality and recalling Lemmata 16 and 18, one easily obtains, proceeding recursively, $F \in \mathrm{Ker}(\mathcal{B}_s)$ (and also, of course, $f \in \mathrm{Ker}(\mathcal{B}_s)$). Furthermore, one can also put F in NF with respect to $\mathcal{A}^+$, i.e. $\mathcal{A}^+(F) = 0$, leaving G in NF with respect to $\mathcal{B}_s$, indeed $[\mathcal{A}^+, \mathcal{B}_s] = 0$ thanks to Lemma 16, and $\mathcal{A}^+ : \mathrm{Ker}(\mathcal{B}_s) \to \mathrm{Ker}(\mathcal{B}_s)$, and the X-JNF is easily obtained. The result on the SJNF follows from Lemma 18. ⊙

If we consider not only two vector fields X_f, X_g, but a whole – abelian or at least nilpotent – algebra $\mathcal{G}$, we have results similar to those obtained for two vector fields. We will first have to generalize, in the obvious way, definitions 2 and 3; i.e. we have:

Definition 2′. *Let us consider a Lie algebra $\mathcal{G}$ of vector fields $\{X_1, ..., X_d\}$, with $X_a = \left(f^{(a)}\right)^i(x)\partial_i$, and let $A^{(a)} = (Df^{(a)})(0)$ be their linear parts. Let these have the S+N decompositions $A^{(a)} = A_s^{(a)} + A_n^{(a)}$. We say that $\mathcal{G}$ is in* Semisimple Joint Normal Form *if and only if all the $f^{(a)}$ are in $\mathcal{K} = \cap_{a=1}^d \text{Ker}\left(\mathcal{A}_s^{(a)}\right)$.*

Notice that – as was the case for vector fields – when the $A^{(a)}$ are semisimple and normal matrices, the definitions of Joint and Semisimple Joint Normal Forms coincide, and algebras which are in Joint Normal Form are also in Semisimple Joint Normal Form.

Definition 3′. *Let us consider a Lie algebra $\mathcal{G}$ of vector fields $\{X_1, ..., X_d\}$, with $X_a = \left(f^{(a)}\right)^i(x)\partial_i$, and let $A^{(a)} = (Df^{(a)})(0)$ be their linear parts. Let these have the S+N decompositions $A^{(a)} = A_s^{(a)} + A_n^{(a)}$. Let the nonlinear parts of $f^{(a)}$ be denoted as $F^{(a)}$, i.e. $f^{(a)}(x) = A^{(a)}x + F^{(a)}(x)$. We say that $\mathcal{G}$ is in X_a-*Joint Normal Form *(where X_a is one of the vector fields spanning the algebra) if and only if, for each fixed a,*

$$F^{(a)} \in \Big(\bigcap_{b\neq a} \text{Ker}(\mathcal{A}_s^{(b)})\Big) \cap \text{Ker}((\mathcal{A}^{(a)})^+)$$

and for all other $f^{(b)}$, $b \neq a$,

$$f^{(b)} \in \bigcap_{b'=1}^{d} \text{Ker}(\mathcal{A}_s^{(b')}) \ .$$

With these, we have the counterpart to Theorem 4 as

Theorem 6. *Let us consider a d-dimensional nilpotent algebra $\mathcal{G}$ of vector fields spanned by X_a, $a = 1, ..., r$, with $X_a = f_i^{(a)}(x)\partial_i$, $(Df^{(a)})(0) = A^{(a)}$, and $f^{(a)}(x) = A^{(a)}x + F^{(a)}(x)$. Let $A^{(a)} = A_s^{(a)} + A_n^{(a)}$ be the S+N decompositions of the matrices $A^{(a)}$. Let $\left(\mathcal{A}^{(a)}\right)^+$ be the homological operator associated with $\left(A^{(a)}\right)^+$, and $\mathcal{A}_s^{(a)}$ be the homological operator associated with $A_s^{(a)}$. Then, by means of a formal series of Poincaré transformations, all the $f^{(a)}$ can be put into Semisimple Joint Normal Form, or, for each fixed a, in X_a-Joint Normal Form.*

Proof. First we note that if the algebra $\mathcal{G}$ is abelian, we can just repeat the proof of Theorem 5, with the role of X played by any of the X_a, obtaining an immediate extension to the case of d-dimensional abelian algebras.

For the nilpotent case the idea is, again, to consider iterated commutation relations: the fact that all semisimple parts $A_s^{(a)}$ and all nilpotent parts $A_n^{(a)}$ are commuting matrices [9,250] reduce to the same discussion as for the abelian case. As already remarked, in the nilpotent case some of the matrices $A_s^{(a)}$ (but not all) are necessarily nilpotent, i.e. $A_s^{(a)} = 0$. Accordingly, for these the result $f^{(a)} \in \text{Ker}(\mathcal{A}_s^{(a)})$ does not produce any restriction.

We can reformulate this in a slightly different way by saying that we consider the chain of algebras $\mathcal{G}_{[k+1]} = [\mathcal{G}, \mathcal{G}_{[k]}]$ with $\mathcal{G}^{[0]} = \mathcal{G}$. If $\mathcal{G}$ is nilpotent, then this ends in $[\mathcal{G}, \mathcal{G}_{[m]}] = 0$, and each factor algebra $\mathcal{G}_{[m]}/\mathcal{G}_{[m+1]}$ is abelian (see the Appendix to this chapter). In particular, $\mathcal{G}_{[m]}$ is abelian. Thus, all the fields in the abelian ideal $\mathcal{G}_{[m]}$ commute with every field. We can then proceed as in the proof of Theorem 4, first for the fields in $\mathcal{G}_{[m]}$ and then, by considering the commutation relations with these, for $\mathcal{G}_{[m-1]}$ and so on, until we have examined all the fields in $\mathcal{G}_{[0]} \equiv \mathcal{G}$. ⊙

Corollary 4. *In any nilpotent* subalgebra $\mathcal{G}_*$ *of a generic algebra* $\mathcal{G}$, *all the nonlinear terms can be put in SJNF or in X-JNF.*

Proof. This just follows from restricting the above normalization procedure to the subalgebra $\mathcal{G}_*$, i.e. by applying the result of Theorem 6 to the appropriate subalgebra. ⊙

Exercise 19. Determine the corresponding matrices A_s and A^+ for the matrix

$$A = \begin{pmatrix} 1 & 0 & 0 \\ 1 & 1 & 0 \\ 0 & 0 & -2 \end{pmatrix} .$$

Show that $x^2 z \in \mathcal{I}_{A_s} \cap \mathcal{I}_{A^+}$. Show that $y^2 z \in \mathcal{I}_{A_s}$ but that it does not belong to $\mathcal{I}_{A^+}$. Does this allow one to make any general statement about the relation of $\mathcal{I}_{A_s}$ and $\mathcal{I}_{A^+}$? Study the homological operators $\mathcal{A}_s = \{A_s x, .\}$ and $\mathcal{A}^+ = \{A^+ x, .\}$ associated with A_s and A^+: Determine $\mathrm{Ker}(\mathcal{A}_s)$ and $\mathrm{Ker}(\mathcal{A}^+)$, and show that $\mathrm{Ker}(\mathcal{A}^+) \subset \mathrm{Ker}(\mathcal{A}_s)$. (Hint: see [96].)

11. The Linearization of a Dynamical System

An application of the JNF deals with the case of *linearizable* DS. Clearly, if $\mathrm{Ker}(\mathcal{A}) = \{0\}$, the DS is linearizable by means of a formal Poincaré transformation. But, whatever the matrix A, the linear VF $X_A = Ax \cdot \nabla$ does clearly commute with the VF S given by

$$S = \sum_i x_i \frac{\partial}{\partial x_i} = (Ix) \cdot \nabla \tag{49}$$

which generates the dilations in R^n. It is easy to see, conversely, that the only VF commuting with S are the linear ones. On the other hand, it is also clear that the identity does not admit resonances. Then we can say [150, 24]:

Corollary 5. *A VF* f *(or a DS* $\dot{x} = f(x)$*) can be linearized if and only if it admits a (possibly formal) symmetry* $g = Bx + G$ *such that* $B = (Dg)(0) = I$.

Remark 3. If the VF f is two-dimensional and $A = (Df)(0)$ has different eigenvalues, and it admits a symmetry $g = Bx + G$ with B not proportional to A, then it can be linearized. Indeed, it is sufficient to consider the other symmetry $af + bg$ such that $aA + bB = I$.

More in general, we can also say:

Corollary 6. *The DS $\dot{x} = f = Ax + F$ can be linearized if and only if it admits a (possibly formal) symmetry $g = Bx + G$ such that B does not admit resonances, i.e.* $\mathrm{Ker}(\mathcal{B}) = \{0\}$. *Actually, to reach this conclusion, it is sufficient that* $\mathrm{Ker}(\mathcal{A}) \cap \mathrm{Ker}(\mathcal{B}) = \{0\}$, *indeed, the JNF of this DS is linear: $\dot{x} = Ax$.*

This result can be easily extended not only to the case of more than one symmetry , as an obvious consequence of Theorem 2, but also to the non-semisimple case [96]. Other comments on this point will be given in Chapter VI, when the convergence of the normalizing transformations will be discussed. Another result, related to the properties of linear DS, is the following [96]:

Theorem 7. *If a DS in R^n can be linearized, then it admits n independent* commuting *symmetries, that can be simultaneously taken into linear form by a coordinate transformation. If, in particular, the system has a diagonalizable A with real eigenvalues, then – once it is linearized and A is diagonal – the dilations $\zeta_i = x_i\partial_i$ (no sum over i) along each direction i, are n linear commuting symmetries for the system. Conversely, if there is a coordinate system where the DS admits n independent linear commuting symmetries $X_i = B_i x$ such that all B_i are semisimple, then the DS can be linearized.*

Proof. In the coordinates where the DS is linear, it is easy to construct n linear commuting symmetries $B_i x \partial_x$, simply by choosing n independent matrices B_i commuting among themselves and with A (if the matrices $I, A, A^2, \ldots, A^{n-1}$ are linearly independent, they immediately provide the matrices B_i; even if this is not the case, the existence of n matrices B_i with the required properties is easily verified if A is put in Jordan form). The existence of the n independent scalings ζ_i in the case of diagonal A is obvious. Conversely, given n semisimple commuting matrices B_i, they can be simultaneously diagonalized: $B_i \to \mathrm{diag}\,(\beta_1^{(i)}, \ldots, \beta_n^{(i)})$. Now, in the basis spanned by the n (independent) vectors $\beta^{(i)} \equiv (\beta_1^{(i)}, \ldots, \beta_n^{(i)})$, the symmetries $\sigma_i = B_i x \partial_x$ become $\sigma_i \to \zeta_i = x_i \partial_i$ (no sum over i), i.e. the independent dilations along the directions $\beta^{(i)}$. A DS admitting such n symmetries is necessarily linear. $\odot$

12. Partial Joint Normal Form for Non-nilpotent Algebras

As already remarked, we cannot expect to obtain for generic Lie algebras results as general as in the case of abelian (or nilpotent) algebras. However, some "Partial" JNF can be obtained, even for non-nilpotent algebras, under some special assumptions. E.g., we have:

Theorem 8. *Let $\mathcal{G}$ be a d-dimensional algebra spanned by $f_a = A_a x + F_a$ ($a = 1, \ldots, d$), and let $\mathcal{G}$ admit a non-trivial center $C(\mathcal{G})$ (e.g., $\mathcal{G}$ is a direct sum $\mathcal{G} = \mathcal{G}_0 \oplus C(\mathcal{G})$). Let the center of $\mathcal{G}$ be spanned by $w_b = C_b x + W_b$ ($b =$*

$1,\dots,d_C$, $d_C \le d$), and assume that the semisimple parts $C_{b,s}$ are normal matrices. Then, by means of formal Poincaré transformations, all fields f_a can be put in the following form

$$f_a \to \widehat{f}_a \in \bigcap_{b=1}^{d_C} \mathrm{Ker}(\mathcal{C}_{b,s}) \qquad (a = 1\dots, d;\ b = 1,\dots,d_C) \tag{50}$$

where, as usual, $\mathcal{C}_{b,s} = \{C_{b,s}x, \cdot\}$. The same result is also true if $\mathcal{G}$ is a direct sum $\mathcal{G}_0 \oplus \mathcal{N}$, where $\mathcal{N}$ is a nilpotent subalgebra.

Proof. According to Theorem 6, there is a formal transformation that takes $w_b \to \widehat{w}_b$ in Semisimple JNF. On the other hand, for all a, b, one has

$$\{f_a, w_b\} = 0$$

which implies (dropping the indices a, b for simplicity) for each order k

$$\mathcal{A}(\widehat{W}_k) - \mathcal{C}(\widetilde{F}_k) = \psi_k \ , \tag{51}$$

where ψ_k is a combination of terms $W_h,\ F_m$ as in the above cases. Applying $\mathcal{C}_s$ to this equation and proceeding inductively, we easily obtain, $\forall k$, $\widetilde{F}_k \in \mathrm{Ker}(\mathcal{C}_s)$, recalling also that $[\mathcal{A}, \mathcal{C}_s] = 0$. ⊙

Other Partial JNF can be obtained in the following *solvable* cases.

Theorem 9. *Let $f = Ax + F,\ g = Bx + G$ satisfy $\{f, g\} = g$. Then, there exists a formal Poincaré transformation which put $F,\ G$ in the following form*

$$F \to \widetilde{F} \in \mathrm{Ker}(\mathcal{B}) \qquad \text{and} \qquad G \to \widehat{G} \in \mathrm{Ker}(\mathcal{B}^+) \tag{52}$$

Proof. By means of formal transformations let us transform G in NF with respect to $\mathcal{B}^+$, i.e. $\widehat{G} \in \mathrm{Ker}(\mathcal{B}^+)$. From $\{f, g\} = g$ we get

$$[A, B] = B \ .$$

It is now easy to see, by choosing a basis in which B is in Jordan form, that B is nilpotent and that A is semisimple with real (and distinct) eigenvalues. In a basis where A is normal, one then gets

$$[A, B^+] = -B^+ \qquad \text{and} \qquad [\mathcal{A}, \mathcal{B}^+] = -\mathcal{B}^+ \ . \tag{53}$$

From $\{f, g\} = g$ one also obtains, with the same notations and procedure as above,

$$\mathcal{A}(\widehat{G}_k) - \mathcal{B}(F_k) = \widehat{G}_k + \psi_k \ . \tag{54}$$

Applying the operator $\mathcal{B}^+$ to this equality and using the above results, in particular (53), one easily gets, proceeding recursively, the partial JNF (52). ⊙

This result can be extended to higher-dimensional solvable algebras of this form:

Theorem 9′. *Let $\mathcal{G}$ be a $(d_a + d_b)$-dimensional algebra spanned by $f_a = A_a x + F_a$ $(a = 1, \dots, d_a)$ and $g_b = B_b x + G_b$ $(b = 1, \dots, d_b)$, and obeying the following commutation rules*

$$\{f_a, f_{a'}\} = \{g_b, g_{b'}\} = 0 \ , \ \{f_a, g_b\} = c_{ab} g_b \quad \text{(no sum over } b) \ ,$$

where c_{ab} are arbitrary constants. Then, for each fixed b, there is a formal Poincaré transformation which takes g_b into the NF form

$$G_b \to \widehat{G}_b \in \operatorname{Ker}(\mathcal{B}_b^+) \tag{55}$$

and all the F_a into

$$F_a \to \widetilde{F}_a \in \operatorname{Ker}(\mathcal{B}_b) \ . \tag{55′}$$

A quite different situation, of special interest in the Hamiltonian case (see Chap. V and also [262]), is covered by the following theorem.

Theorem 10. *Assume that the given DS $\dot{x} = f(x)$ admits a d-dimensional solvable algebra of symmetries $\mathcal{G} = \{g_1, \dots, g_d\}$:*

$$\{f, g_i\} = 0 \qquad g_i = B_i x + G_i \qquad (i = 1, \dots, d)$$

Then, by means of a formal Poincaré transformation we can obtain the following NF:

$$\begin{aligned} f &\to \widehat{f} \in \Big(\bigcap_{i=1}^{d} \operatorname{Ker}(\mathcal{B}_{i,s}) \Big) \cap \operatorname{Ker}(\mathcal{A}^+) \\ g_i &\to \widetilde{g}_i \in \operatorname{Ker}(\mathcal{A}_s) \end{aligned} \tag{56}$$

Proof. Solvability of $\mathcal{G}$ means that

$$\{\mathcal{G}, \mathcal{G}\} = \mathcal{G}^{(1)}, \quad \{\mathcal{G}^{(1)}, \mathcal{G}^{(1)}\} = \mathcal{G}^{(2)}, \quad \dots, \quad \{\mathcal{G}^{(r)}, \mathcal{G}^{(r)}\} = 0 \ .$$

Now, we can first put each element $g_r \in \mathcal{G}^{(r)}$ in $\operatorname{Ker}(\mathcal{B}_{r,s})$ and also in the intersection over r of $\operatorname{Ker}(\mathcal{B}_{r,s})$ (with clear notation), as in Theorem 6 ($\mathcal{G}^{(r)}$ is in fact abelian). As already noted, some of the B_i are nilpotent, thus $B_{i,s} = 0$ and in this case clearly no transformation is required. Using the same arguments as for previous theorems, we can conclude also that

$$f \to f \in \operatorname{Ker}(\mathcal{B}_{r,s}) \ .$$

Next, we can transform this f into

$$f \to f \in \operatorname{Ker}(\mathcal{B}_{(r-1),s})$$

by using $\{G^{(r-1)}, \mathcal{G}^{(r)}\} = \mathcal{G}^{(r)}$ and the same procedure as in the proof of Theorem 9. This procedure can be repeated for all g_i, $i = 1, \dots, d$. Finally, another Poincaré transformation can take f, as obtained in this way, into NF with respect to A^+. The final result (56) is obtained in the usual way, and by observing that at each step the normality of f is preserved. ⊙

Examples for all the above JNF will be provided in Chapter V, in the context of Hamiltonian problems.

Appendix. Some Results on Matrices and Lie Algebras

The purpose of this appendix is to collect some known results of algebraic nature that are relevant to the discussion of this chapter and which are, explicitly or implicitly, used here. These are provided only for the convenience of the reader, but they should be well known. Thus, this section should be skipped and used only as a reference in case of doubt. For proofs and more details, the reader is referred e.g. to [203,250], which we follow closely in what follows.

First of all, we recall some basic definitions and properties concerning a single (real) matrix A. We will then go on to consider Lie algebras. In doing this, we will denote by $[\mathcal{F},\mathcal{G}]$ the set of elements of the form $[f,g]$ with $f \in \mathcal{F}$, $g \in \mathcal{G}$.

1. We say that A is *normal* if it commutes with its adjoint A^+, i.e. if

$$\left[A, A^+\right] = 0 \ . \tag{A.1}$$

Thus, a selfadjoint matrix is also normal. In the case of real matrices, $A^+ = A^T$, and a symmetric matrix is also normal.

2. We say that A is *semisimple* if it is linearly conjugate to a diagonal matrix, i.e. if there exists a matrix U such that

$$UAU^{-1} = D \quad , \qquad D_{ij} = \lambda_i \, \delta_{ij} \ . \tag{A.2}$$

Notice that the diagonal matrix D is surely normal, but the semisimple matrix A is not necessarily normal; we are guaranteed this to be the case only if the matrix U is orthogonal, i.e. if $U^+ = U^{-1}$.

3. If two semisimple matrices A and B commute, they can be taken into diagonal form by the same transformation, i.e. by the same matrix U.

4. We say that A is *nilpotent* if there is a finite integer p such that $A^p = 0$; the integer p is also called the degree of nilpotency of A.

5. Any matrix A can be decomposed as $A = A_s + A_n$, where A_s is semisimple, A_n is nilpotent, and $[A_s, A_n] = 0$. If A is real, then A_s and A_n can be chosen to be real. The matrices A_s and A_n can be written as polynomials in A, and any matrix B which commutes with A also commutes with A_s and with A_n.

6. If the matrices A_i, with $i = 1, ..., r$, form a set that is closed under commutation, i.e. if there are real numbers c_{ij}^k such that

$$[A_i, A_j] \ = \ c_{ij}^k \ A_k \ , \tag{A.3}$$

we say that the A_i form a *Lie algebra*; the c_{ij}^k are also called structure constants for the Lie algebra. By exponentiating the action of a Lie algebra $\mathcal{G}$, we get a Lie group G.

7. A Lie algebra has the structure of a linear space. Its dimension as a linear space is also called the dimension of the Lie algebra. A set of elements of the algebra that provide a basis for the Lie algebra as a linear space is also called a set of *generators* of the algebra. A Lie algebra is better thought of in abstract terms as a set $\mathcal{G}$ of objects g_i equipped with an antisymmetric binary operation

$[.,.] : \mathcal{G} \times \mathcal{G} \to \mathcal{G}$. Any realization of this structure in terms of concrete objects (matrices, operators...) is a *representation* of the Lie algebra.

8. For any Lie algebra, we can consider its *adjoint representation*. This is obtained by associating with each element $g \in \mathcal{G}$ the linear operator $\mathbf{ad}_g : \mathcal{G} \to \mathcal{G}$ defined as $\mathbf{ad}_g : h \to [g, h]$. This can be naturally mapped into a set of matrices of the same dimension n as the Lie algebra itself.

9. A subset $\mathcal{F} \in \mathcal{G}$ of a Lie algebra is a *Lie subalgebra* if $[\mathcal{F}, \mathcal{F}] \subseteq \mathcal{F}$. By exponentiating a subalgebra $\mathcal{F}$ of $\mathcal{G}$, we get a Lie subgroup F of G.

10. A subset $\mathcal{I} \in \mathcal{G}$ of a Lie algebra is an *ideal* if $[\mathcal{G}, \mathcal{I}] \subseteq \mathcal{I}$. Clearly, an ideal is also a subalgebra. If $\mathcal{I}$ is an ideal in $\mathcal{G}$, the factor algebra $\mathcal{H} = \mathcal{G}/\mathcal{I}$ is also a Lie algebra. The corresponding Lie subgroup $I \subseteq G$ is a normal subgroup of G, and $H = G/I$ is a Lie group. For any Lie algebra $\mathcal{G}$, $\mathcal{G}$ itself and $\{0\}$ are ideals; we call these *trivial*.

11. If $\mathcal{I}$ and $\mathcal{H}$ are ideals in $\mathcal{G}$, then $[\mathcal{I}, \mathcal{H}]$ is also an ideal in $\mathcal{G}$.

12. The set of elements commuting with the whole Lie algebra $\mathcal{G}$ is called the *center* of $\mathcal{G}$, and denoted $\mathcal{C}(\mathcal{G})$ (or $\mathcal{C}$ for short). Clearly, $[\mathcal{G}, \mathcal{C}] = 0$, and $\mathcal{C}$ is an abelian ideal in $\mathcal{G}$. The corresponding Lie subgroup C is the center of G.

13. The centralizer of a subalgebra $\mathcal{F}$ in $\mathcal{G}$ is the greatest subalgebra $\mathcal{H} \subseteq \mathcal{G}$ such that $\mathcal{C}(\mathcal{H}) = \mathcal{F}$.

14. For any Lie algebra $\mathcal{G}$, we can define a sequence of Lie algebras $\{\mathcal{G}_0, \mathcal{G}_1, ...\}$ by

$$\mathcal{G}_{n+1} = [\mathcal{G}, \mathcal{G}_n] \ , \tag{A.4}$$

with $\mathcal{G}_0 = \mathcal{G}$. If this sequence terminates in zero, i.e. if $\mathcal{G}_n = \{0\}$ for $n \geq p$, we say that the Lie algebra is *nilpotent* of degree (or rank) p.

15. For any Lie algebra $\mathcal{G}$, we can define a sequence of Lie algebras $\{\mathcal{G}^0, \mathcal{G}^1, ...\}$ by

$$\mathcal{G}^{n+1} = [\mathcal{G}^n, \mathcal{G}^n] \ , \tag{A.5}$$

with $\mathcal{G}^0 = \mathcal{G}$. If this sequence terminates in zero, i.e. if $\mathcal{G}^n = \{0\}$ for $n \geq p$, we say that the Lie algebra is *solvable* of degree (or rank) p.

16. Any Lie subalgebra of a nilpotent Lie algebra is nilpotent; any Lie subalgebra of a solvable Lie algebra is solvable.

17. The sequence of the $\mathcal{G}_n$ is called the *descending central series* of $\mathcal{G}$; the sequence of the $\mathcal{G}^n$ is called the *derived series* of $\mathcal{G}$; the Lie algebra $\mathcal{G}^1$ is called the *derived algebra* – or the *derived ideal* – of $\mathcal{G}$.

18. For each pair of algebras $\mathcal{G}_r, \mathcal{G}_s$ with $0 \leq r \leq s$, $\mathcal{G}_s$ is an ideal in $\mathcal{G}_r$. Similarly, for each pair of algebras $\mathcal{G}^r, \mathcal{G}^s$ with $0 \leq r \leq s$, $\mathcal{G}^s$ is an ideal in $\mathcal{G}^r$.

19. Each factor algebra $\mathcal{G}^n/\mathcal{G}^{n+1}$ is abelian. If $\mathcal{I}$ is an ideal in $\mathcal{G}^n$ and $\mathcal{G}^n/\mathcal{I}$ is abelian, then $\mathcal{G}^{n+1} \subseteq \mathcal{I}$.

20. Each factor algebra $\mathcal{G}_n/\mathcal{G}_{n+1}$ is abelian.

21. Clearly, from the above definitions, $\mathcal{G}^1 = \mathcal{G}_1$, and $\mathcal{G}^n \subseteq \mathcal{G}_n$ for all n. Thus, any nilpotent Lie algebra is also solvable.

22. If the Lie algebra $\mathcal{G}$ is solvable, then there exists a (generally, complex) matrix representation of $\mathcal{G}$ in terms of upper (lower) triangular matrices, i.e. of matrices A such that $A_{ij} = 0$ for $i > j$ (for $i < j$).

23. If $\mathcal{G}$ is solvable, then $\mathcal{G}_1 = \mathcal{G}^1$ is nilpotent. Conversely, if $\mathcal{G}_1 = \mathcal{G}^1$ is nilpotent, then $\mathcal{G}$ is solvable.

24. We say that the Lie algebra $\mathcal{G}$ is *simple* if does not have nontrivial ideals. We say that it is *semisimple* if it does not have abelian nontrivial ideals. Any simple Lie algebra is also semisimple.

25. For the Lie algebra $\mathcal{G}$, there exists a solvable ideal $\mathcal{R}$ that contains any solvable ideal of $\mathcal{G}$. The ideal $\mathcal{R}$ is also called the *radical* of $\mathcal{G}$. For the Lie algebra $\mathcal{G}$, there exists a nilpotent ideal $\mathcal{N}$ which contains any nilpotent ideal of $\mathcal{G}$. The ideal $\mathcal{N}$ is also called the *nilradical* of $\mathcal{G}$. A differentiation of $\mathcal{G}$ maps $\mathcal{R}$ into $\mathcal{N}$.

26. The radical $\mathcal{R}$ of a Lie algebra $\mathcal{G}$ is the smallest of the ideals $\mathcal{I}$ for which the factor algebra $\mathcal{G}/\mathcal{I}$ is semisimple.

27. For the Lie algebra $\mathcal{G}$ with radical $\mathcal{R}$ there exists a semisimple Lie algebra $\mathcal{S}$ such that $\mathcal{G} = \mathcal{S} \oplus \mathcal{R}$. Any two such semisimple algebras are conjugated by an automorphism of $\mathcal{G}$ that leaves $\mathcal{R}$ globally invariant.

28. The semisimple Lie algebras can be completely classified. Any semisimple Lie algebra is the direct sum of semisimple ideals.

29. The solvable Lie algebras are not classified. Such a classification would require to solve the problem of simultaneous reduction to canonical form of n matrices, that is not solved, even for $n = 2$ [203].

30. The *Killing form* $K : \mathcal{G} \times \mathcal{G} \to \mathbf{R}$ is defined as

$$K(g,h) = \mathrm{Tr}\left[\mathrm{ad}_g \mathrm{ad}_h\right] . \tag{A.6}$$

This is invariant under automorphisms of the Lie algebra.

31. The Lie algebra $\mathcal{G}$ is nilpotent if and only if $K(\mathcal{G},\mathcal{G})$ is identically zero. The Lie algebra $\mathcal{G}$ is semisimple if and only if $K(\mathcal{G},\mathcal{G})$ is nonsingular.

32. The Lie algebra $\mathcal{G}$ is solvable if and only if $K(\mathcal{G},[\mathcal{G},\mathcal{G}])$ – which is just $K(\mathcal{G},\mathcal{G}_1)$ – is identically zero.

33. The Lie algebra $\mathcal{G}$ is nilpotent if and only if ad_g is nilpotent for all $g \in \mathcal{G}$.

34. If $\mathcal{G}$ is nilpotent, then its center $C(\mathcal{G})$ is nonzero.

35. If the matrices A_i generate a nilpotent Lie algebra of rank p, then the product of any p such matrices vanishes.

V. Normal Forms and Symmetries for Hamiltonian Systems

In the previous Chap. IV we have analyzed Poincaré-Dulac normal forms in the presence of nonlinear symmetries. In the present chapter, we show how this theory extends to the framework of Hamiltonian systems. Thus, we will be considering a Hamiltonian H which admits a symmetry S, and see how this symmetry property influences the Birkhoff-Gustavson normal form of H.

We will only directly discuss the Hamiltonian case, without showing how these same results could be extracted from the case of Poincaré-Dulac normal forms for general first order ODEs by restriction to the Hamiltonian case. The latter approach would give an NF expansion for the Hamilton equations of motion, while working directly in the Hamiltonian framework. We will rather consider the NF expansion for the Hamiltonian H itself [13,17,68,174,209,210,246–248,317].

Actually, our proofs here will go essentially along the same lines – and are based on the same ideas – as those for the Poincaré-Dulac case. However, formulating these results in the Hamiltonian language is especially useful in view of the applications to concrete Hamiltonian systems.

1. Birkhoff-Gustavson Normal Forms

The Birkhoff-Gustavson normal forms have been introduced in Sect. 6 of Chap. II; here we quickly recall some fundamental notions and notations.

Let $M \subseteq \mathbf{R}^\ell$ be a smooth manifold, and let (q,p) be coordinates in T_*M. For notational convenience, we will also consider coordinates $x_i \in \mathrm{T}_*M \subseteq \mathbf{R}^{(2\ell)} \equiv \mathbf{R}^n$ given by $x_i = q_i$ for $1 \le i \le \ell$ and $x_i = p_i$ for $\ell < i \le 2\ell$.

We will denote by $\mathcal{F}$ the space of functions $F : \mathrm{T}_*M \to \mathbf{R}$ (with vanishing zeroth and first-order terms), and by $\mathcal{F}_k \subset \mathcal{F}$ the space of such functions which are homogeneous of degree $k+2$.

A Hamiltonian is a function $H \in \mathcal{F}$. We will write it as

$$H = H_0 + H_R \ , \tag{1}$$

where H_0 represents the quadratic part, and H_R the higher-order terms. We also write

$$H(q,p) \ = \ \sum_{k=0}^{\infty} \ H_k(q,p) \ . \tag{2}$$

The (antisymmetric) Poisson bracket $\{.,.\} : \mathcal{F} \times \mathcal{F} \to \mathcal{F}$, already considered in previous chapters, is defined as

$$\{F, G\} = \sum_{i=1}^{\ell} \left[\frac{\partial F}{\partial q_i} \frac{\partial G}{\partial p_i} - \frac{\partial F}{\partial p_i} \frac{\partial G}{\partial q_i} \right] . \tag{3}$$

As stressed in Chap. II, $\{H_0, .\} : \mathcal{F}_k \to \mathcal{F}_k$. We denote this operator as

$$\zeta := \{H_0, .\} . \tag{4}$$

To any $H \in \mathcal{F}$ we associate the vector field X_H on T_*M generated by H according to the Hamilton equations (Chap. II). The time evolution of any $F \in \mathcal{F}$ along the dynamics generated by H is given simply by $X_H(F)$, and we have $X_H(F) = \{F, H\}$, so $\zeta = -X_{H_0}$. We denote by $\mathcal{M}$ the set of vector fields, compatible with the (standard) symplectic structure, on T_*M.

Let us now come back to the Hamiltonian H, and let us consider a change of coordinates[1]

$$q = Q + \partial F / \partial P \ ; \ p = P - \partial F / \partial Q \tag{5}$$

with $F \in \mathcal{F}_k$. This is called a Birkhoff (B) transformation.

Under this, the Hamiltonian H is changed to $\widetilde{H} = L_F(H)$, where the terms $\widetilde{H}_m$ with $m < k$ are unchanged, and $\widetilde{H}_k$ is given by

$$\widetilde{H}_k = H_k - \{H_0, F_k\} = H_k - \zeta(F_k) . \tag{6}$$

In this way, we can eliminate all the terms in Ran(ζ). If we introduce in $\mathcal{F}$ a scalar product analogous to the one introduced in $\mathcal{V}$ (i.e. the standard scalar product of Chaps. II and IV), we have that H can be brought by a series of B transformations into

$$\widehat{H} = H_0 + \sum_{k=1}^{\infty} \widehat{H}_k , \tag{7}$$

where all the $\widehat{H}_k$ satisfy

$$\widehat{H}_k \in [\mathrm{Ran}(\zeta)]^{\perp} = \mathrm{Ker}\left(\zeta^+\right) . \tag{8}$$

As in the case of Poincaré normal forms, we will assume that the relevant homological operator, i.e. ζ, is normal. Thus, with the adjoint defined in terms of our standard scalar product, we will have $[\zeta, \zeta^+] = 0$.

The reduction of H to its normal form $\widehat{H}$ as in (7) requires us to use, at each step, the appropriate F_m. This is obtained as the solution of the *homological equation*

$$\zeta(F_m) = \pi_\zeta H_m , \tag{9}$$

where we denote the projection $\pi_\zeta : \mathcal{M} \to \mathrm{Ran}(\zeta)$ by π_ζ. The solutions to these are determined up to terms in Ker(ζ).

A system in the form of (7) that satisfies (8) is said to be in *Birkhoff-Gustavson normal form*.

---

[1] See Chap. II for the connection with the case of generic dynamic systems and with Lie operators [47].

Definition 1. *If $H \in \mathcal{F}$ is a Hamiltonian, and the function $S \in \mathcal{F}$ is such that $\{H, S\} = 0$, we say that S is a symmetry of H.*

Lemma 1. *If S is a symmetry of H, then $[X_H, X_S] = 0$.*

Proof. As already remarked in Chap. II, $\{H, S\} = 0$ implies $[X_H, X_S] = 0$. The latter condition, however, only implies $\{H, S\} = c$. ⊙

Notice that the converse of Lemma 1 does not hold (see also Exercise 3), i.e. we could have vector fields X commuting with X_H but which cannot be expressed in terms of a function S. Thus, the symmetries we consider here are the "Hamiltonian" ones [262], and correspond to the constants of motion for the field X_H.

Thus we have that the Birkhoff-Gustavson normal form condition (Chap. II), can be interpreted as a condition on the symmetry of H; indeed, we can rephrase the Birkhoff theorem as:

Theorem 1. *The Hamiltonian $H = H_0 + H_R$ is in Birkhoff-Gustavson normal form if and only if $H_R \in \text{Ker}(\zeta^+)$, where $\zeta = \{H_0, .\}$ is the homological operator associated with H_0. If ζ is normal, then $H = H_0 + H_R$ is in normal form if and only if H_0 is a symmetry of H, i.e. $\{H_0, H\} = 0$. Any H can be formally taken into normal form by a formal series of Birkhoff transformations.*

In the following, we will be most concerned with the symmetry of H. Thus, we will collect here some simple results concerning these. These will parallel results concerning the symmetry of vector fields. Indeed, as Lemma 1 shows, the commutation of functions in $\mathcal{F}$ (under the Poisson bracket) entails the commutation of the corresponding vector fields on T_*M.

We denote by $\mathcal{G}_H \subset \mathcal{F}$ the set of functions that are symmetries of a given Hamiltonian H. This is a ring, and in particular an abelian algebra under the sum. However, we will be more interested in its Lie algebra structure. We have indeed, by the Jacobi identity and Lemma 1:

Lemma 2. *The set $\mathcal{G}_H$, equipped with the Poisson bracket, is a Lie algebra.*

In the following, we will work directly in $\mathcal{F}$ – i.e. on the Hamiltonians – rather than consider the associated vector fields in $\mathcal{M}$. Exercises 1-4 will help to clarify the slightly different algebraic structures one is faced with when considering symmetries in the two settings.

We also have the two following corollaries:

Corollary 1. *All terms in $\text{Ker}(\zeta)$, and all terms $\widehat{H}_m$ when the Hamiltonian is in normal form, are symmetries (or constants of motion) of the quadratic part H_0, and the quadratic part H_0 is a symmetry of the full Hamiltonian $\widehat{H}$ when this is in normal form.*

Corollary 2. *Any Hamiltonian in normal form $\widehat{H}$ with $\ell = 2$ degrees of freedom is (Liouville) integrable. Indeed, either $\widehat{H}$ is a function of H_0 only, or H and H_0 are two independent constant of motion.*

See [187] for an example of a Hamiltonian in NF and in $\ell = 3$ degrees of freedom which is *not* integrable.

Exercise 1. Consider the set of vector fields $X_S \in \mathcal{M}$ associated with functions $S \in \mathcal{G}_H$, equipped with the standard commutator. Show that this is a Lie algebra.

Exercise 2. Consider the set $\mathcal{I}_H$ of the constants of motion under X_H. These will naturally be the functions $f \in \mathcal{F}$ such that $X_H(f) = 0$. Show that these are just the symmetries of H.

Exercise 3. Consider again the vector field $X_H \in \mathcal{M}$, and denote by $\mathcal{A}_H \subset \mathcal{M}$ its symmetry algebra, i.e. $\mathcal{A}_H = \{Y \in \mathcal{M} \ : \ [X_H, Y] = 0\}$. What relation exists between $\mathcal{A}_H$ and $\mathcal{G}_H$? Show that $\mathcal{A}_H$ is a module over the algebra $\mathcal{I}_H$ defined in Exercise 2.

Exercise 4. Does the set $\mathcal{G}_H$ have the structure of a module, and if yes, over which algebra ?

Exercise 5. Consider the Hamiltonians (for $M = \mathbf{R}^2$) $K = (1/2)(q_1^2 + p_1^2) + (q_1 q_2 + p_1 p_2)^2$, and $H = (1/2)(q_1^2 + q_2^2 + p_1^2 + p_2^2) + K^2$. Are these in normal form?

Exercise 6. Consider the Hamiltonians K, H introduced in Exercise 5. Denote by K_0, H_0 their quadratic parts. Compute $\{H, K\}$, $\{H_0, K\}$, and $\{K_0, H\}$.

2. Birkhoff-Gustavson Normal Forms with Symmetries

Let us now consider the case when H admits a nontrivial[2] function $S \in \mathcal{F}$ such that

$$\{H, S\} = 0 \ . \tag{10}$$

As remarked above, this implies in particular that

$$[X_H, X_S] = 0 \ . \tag{11}$$

The geometrical relation (11), stating the commutation of the flows under X_H and under X_S, does not depend on the coordinate representation, and is therefore invariant under the change of coordinates needed to transform H into Birkhoff-Gustavson normal form. However, it is clear that the coordinate expression for S will be changed in this transformation.

Let us consider, together with the expansion (2) for H, the analogous expansion[3] for S:

$$S = S_0 + \sum_{k=1}^{\infty} S_k \qquad (S_k \in \mathcal{F}_k) \ . \tag{12}$$

[2] Here, by nontrivial we mean that S is not a function of H, including the case $S = c$.

[3] Notice that we are implicitly excluding consideration of symmetry functions S with a linear part.

Rewriting S and H as their power expansion, and collecting terms of the same order, we get

$$\{H,S\} = \{H_0,S_0\} + \sum_{k=1}^{\infty}\left[\sum_{r=0}^{k}\{H_r,S_{k-r}\}\right] . \tag{13}$$

The first term on the right has to vanish – as it is the only one in $\mathcal{F}_0$ – i.e. (10) implies

$$\{H_0,S_0\} = 0 . \tag{14}$$

In the language of symmetry algebras, (14) means $S \in \mathcal{G}_H$ implies $S_0 \in \mathcal{G}_{H_0}$; compare this with results in Chap. IV.

As for the k sum in (13), each of the terms in the square bracket has to vanish separately, which implies – pointing out explicitly the role of H_0 and S_0 – that we have for any $k \geq 1$ the equation

$$\{H_0,S_k\} - \{S_0,H_k\} + \sum_{r=1}^{k-1}\{H_r,S_{k-r}\} = 0 . \tag{15}$$

We will introduce the (homological) operator

$$\omega(.) = \{S_0, .\} \tag{16}$$

analogous to ζ; with this, (14) reads

$$\zeta(S_0) = \omega(H_0) = 0 \tag{17}$$

and (15) is rewritten, with $k > 1$, as

$$\zeta(S_k) - \omega(H_k) = -\sum_{r=1}^{k-1}\{H_r,S_{k-r}\} \equiv \psi_k . \tag{18}$$

Let us first consider, to fix ideas, the case $S = S_0$. Now (18), or equivalently (10), reduces to $\{H_k,S_0\} = 0$, i.e.

$$\omega(H_k) = 0 \quad \forall k \geq 0 , \tag{19}$$

which of course also means

$$H \in \mathrm{Ker}(\omega) . \tag{20}$$

When we take H into NF, in the normalization process H could, a priori, be changed into something not contained in $\mathrm{Ker}(\omega)$. However, the fact that the normalizing transformation amounts to a Poisson bracket operation (which is also the case for the ω action) and the Jacobi identity ensure that this is not the case; i.e. we have the following theorems:

Theorem 2. *If H admits a quadratic symmetry $S_0 \in \mathcal{F}_0$, then H can be reduced to the Birkhoff-Gustavson Normal Form $\widehat{H} = \sum_{r=0}^{\infty}\widehat{H}_r$, where $\widehat{H}_0 = H_0$, and $\widehat{H}_r \in \mathrm{Ker}(\zeta) \cap \mathrm{Ker}(\omega)$, where $\zeta = \{H_0, .\}$, $\omega = \{S_0, .\}$. Then, in particular, $\widehat{H}$ also admits the symmetry S_0.*

Theorem 3. *If H admits the quadratic symmetry $S_0 \in \mathcal{F}_0$, then S_0 is also a symmetry of its quadratic part H_0.*

We point out that Theorem 3, which is actually just a reformulation of (14) and (17) above, is very useful[4] in the determination of actual symmetries of H: We can first determine symmetries of H_0, and then consider only these in order to find solutions to $\{H, S\} = 0$. In Chap. IV we discuss the similar case encountered in Poincaré-Dulac theory.

Let us now consider the general case, i.e. the case when S is a generic power series (starting with quadratic terms). We will need two simple lemmas, which we state in the notation used so far, and whose proofs amount to the Jacobi identity:

Lemma 3. *If $\{H_0, S_0\} = 0$, then $[\zeta, \omega] = 0$. More generally, $[\zeta, \omega](.) = \{\{H_0, S_0\}, .\}$.*

Lemma 4. *If we write $\zeta_A(.) = \{A, .\}$ for the homological operator associated to a quadratic function $A \in \mathcal{F}_0$, then $\mathrm{Ker}(\zeta_A)$ is invariant under the Poisson bracket, i.e. $\{.,.\} : \mathrm{Ker}(\zeta_A) \times \mathrm{Ker}(\zeta_A) \to \mathrm{Ker}(\zeta_A)$.*

Proof. Indeed, by the Jacobi identity we have $\zeta_A(\{f, g\}) = \{f, \{A, g\}\} - \{g, \{A, f\}\}$, and this vanishes for f and g in $\mathrm{Ker}(\zeta_A)$. ⊙

With these, we are able to prove

Theorem 4. *Let H, S be given by*

$$H = H_0 + \sum_{k=1}^{\infty} H_k \quad ; \quad S = S_0 + \sum_{k=1}^{\infty} S_k \ ,$$

(where H_k, S_k are in $\mathcal{F}_k$). Let $\{H, S\} = 0$, and let $\zeta = \{H_0, .\}$ and $\omega = \{S_0, .\}$ be normal operators. By applying the Birkhoff normalizing transformation, these can be changed respectively to

$$\widehat{H} = H_0 + \sum_{k=1}^{\infty} \widehat{H}_k \quad ; \quad \widetilde{S} = S_0 + \sum_{k=1}^{\infty} \widetilde{S}_k$$

where $\{\widehat{H}, \widetilde{S}\} = 0$, and

$$\widehat{H}_k \in \mathrm{Ker}(\zeta) \quad , \quad \widetilde{S}_k \in \mathrm{Ker}(\zeta) \ .$$

Proof. Let us come back to (13), and suppose we first take H into NF. Let us first consider $k = 1$, in which case (15) reads simply

$$\zeta(S_1) = \omega(H_1) \ . \tag{21}$$

Now, let us apply ζ on both sides of (21), and make use of the corollary above:

[4] This is the only reason for restating (14) as a theorem.

$$\zeta\left(\zeta(S_1)\right) = \zeta\left(\omega(H_1)\right) = \omega\left(\zeta(H_1)\right) \ , \tag{22}$$

and if H is in Birkhoff-Gustavson NF, the r.h.s. term vanishes. Thus,

$$\zeta^2(S_1) = 0 \ . \tag{23}$$

The normality[5] of ζ entails

$$\mathcal{F} = \mathrm{Ran}(\zeta) \oplus \mathrm{Ker}(\zeta) \ , \tag{24}$$

and therefore (23) actually implies

$$\zeta(S_1) = 0 \ . \tag{25}$$

Now, let us look again at (15). It is clear that, if $S_m \in \mathrm{Ker}(\zeta) \quad \forall m \leq k-1$, and H is in Birkhoff-Gustavson NF (up to order k at least) the terms in the sum that defines ψ_k [see (18)] are all in $\mathrm{Ker}(\zeta)$, see Lemma 4.

Thus, applying ζ to (15) and using $\zeta\left(\psi_k\right) = 0$, we get

$$\zeta\left(\zeta(S_k)\right) = \zeta\left(\omega(H_k)\right) = \omega\left(\zeta(H_k)\right) = 0 \ , \tag{26}$$

where the last equality follows from the assumption that H is in Birkhoff-Gustavson NF at least up to order k. This means, exactly in the same way as in the $k = 1$ case, that $\zeta(S_k) = 0$, i.e. $S_k \in \mathrm{Ker}(\zeta)$. This argument can be repeated for all the k, and the proof is thus completed. ⊙

Clearly, the symmetry of roles between the H and the S is broken by the fact that we put H – and not S – in normal form. Obviously, we could proceed the other way around, and normalize S: In this case, we would get $\widehat{S}$ and $\widetilde{H}$ satisfying $\{\widetilde{H}, \widehat{S}\} = 0$, with $\widetilde{H}$ and $\widehat{S}$ in $\mathrm{Ker}(\omega)$. However, as we will see in a moment, the two can actually be put in a joint normal form (JNF) that restores this symmetry.

Definition 2. *The functions $H, S \in \mathcal{F}$ are in JNF if they are both in* $\mathrm{Ker}(\zeta) \cap \mathrm{Ker}(\omega)$, *where $\zeta = \{H_0, .\}$, $\omega = \{S_0, .\}$, and H_0, S_0 are the quadratic parts of H, S.*

Theorem 5. *Let H, S be given by*

$$H = H_0 + \sum_{k=1}^{\infty} H_k \ \ ; \ \ S = S_0 + \sum_{k=1}^{\infty} S_k \ ,$$

where H_k, S_k are in $\mathcal{F}_k$. Let $\{H, S\} = 0$ and let both $\zeta = \{H_0, .\}$ and $\omega = \{S_0, .\}$ be normal. Then, H and S can be transformed in JNF by a sequence of Birkhoff transformations.

[5] We recall that here we have assumed that ζ is normal. The conditions on A, i.e. on H_0, which correspond to this case are discussed in the final section of this chapter.

Proof. As pointed out above, the normalizing transformations needed to take H to its Birkhoff-Gustavson NF, i.e. L, are only defined up to $\delta F_m \in \mathrm{Ker}(\zeta)$. However, in the change of coordinates the change in the S_k depends on ω, not on ζ; more precisely, S_m is transformed into $\widetilde{S}_m$, given by

$$\widetilde{S}_m = S_m - \omega(F_m) \ . \tag{27}$$

Thus, after choosing the F_m to solve the homological equation (9) for ζ, we can still consider a Birkhoff transformation with $\delta F_m \in \mathrm{Ker}(\zeta)$. This will not change the H_m, but the S_m will be changed to

$$\widetilde{S}_m = S_m - \omega(\delta F_m) \ . \tag{28}$$

Thus, we can eliminate in this way terms in the image of $\mathrm{Ker}(\zeta)$ under ω. The corollary to Lemma 3 above immediately implies that $\omega : \mathrm{Ker}(\zeta) \to \mathrm{Ker}(\zeta)$; as we assumed that ω is normal, $\mathrm{Ran}(\omega) = [\mathrm{Ker}(\omega)]^\perp$, and

$$\mathcal{F} = \mathrm{Ker}(\omega) \oplus \mathrm{Ran}(\omega) \ . \tag{29}$$

With this, we can eliminate terms in

$$\mathrm{Ker}(\zeta) \cap \mathrm{Ran}(\omega) \ , \tag{30}$$

and by choosing appropriately δF_m we can obtain

$$\widetilde{S}_m \in \mathrm{Ker}(\zeta) \cap \mathrm{Ker}(\omega) \ . \tag{31}$$

In explicit terms, let π_ω be the projection

$$\pi_\omega : \mathcal{F} \to \mathrm{Ran}(\omega) \ . \tag{32}$$

The δF_m should then be chosen to be the solution to the homological equation

$$\omega(\delta F_m) = \pi_\omega(S_m) \ . \tag{33}$$

It is quite clear that we do not need to consider two successive Birkhoff transformations. We can equally well consider a unique B transformation of generator $F_m = t_m + \vartheta_m$, where t_m and $\vartheta_m \in \mathrm{Ker}(\zeta)$ are chosen as solutions to the pair of homological equations (obviously the ϑ_m in the first of these is redundant, and has been inserted only to preserve the formal symmetry of the equations)

$$\begin{cases} \zeta(t_m + \vartheta_m) = \pi(H_m) \\ \omega(t_m + \vartheta_m) = \sigma(S_m) \end{cases} .$$

The proof is now complete. ⊙

Some simple but relevant consequences of this result are contained in the following Corollaries and Remarks.

Corollary 3. *If the Hamiltonian H admits a symmetry S, as in Theorem 4, then the quadratic part S_0 of S is a symmetry of the quadratic part H_0 of H. Once H is in NF, i.e. $H \to \widehat{H}$ and $S \to \widetilde{S}$, then $\widetilde{S}$ is a symmetry of the* *quadratic part H_0 of H. In other words, S_0 and $\widetilde{S}$ are constants of motion of H_0. Conversely, if S is put in NF, $S \to \widehat{S}$, then $\widetilde{H}$ is a constant of motion of S_0. If H and S are in JNF, $\widehat{H}$, $\widehat{S}$, as in Theorem 5, then all terms $\widehat{H}_k$ and $\widehat{S}_m$ (and $\widehat{H}$, $\widehat{S}$ as well) are simultaneous constants of motion of the problems having the quadratic functions H_0 and S_0 as Hamiltonians or, equivalently, of the two linear problems $\dot{x} = J\nabla H_0$ and $\dot{x} = J\nabla S_0$).*

Remark 1. If H_0 has the form (which is in fact the "standard Birkhoff form")

$$H_0 = \frac{1}{2}\sum_{r=1}^{\ell} \omega_r \left(q_r^2 + p_r^2\right) \qquad (\omega_r > 0) \tag{34}$$

and the frequencies ω_r are rationally dependent, i.e.

$$\sum_{r=1}^{\ell} k_r \omega_r = 0 \qquad k_r \in Z$$

it is customary to say that the frequencies are resonant. It should be completely clear that this notion of resonance is somewhat different from the notion of resonant eigenvalues of a DS in the Poincaré-Dulac sense: resonant frequencies actually lead to resonant terms in the PD sense, but the converse is not true. Precisely, when the frequencies in (34) are not resonant, the PD resonant terms in the associated DS $\dot{x} = J\nabla H_0$, or, which is the same, the terms in Ker(ζ), are just the functions of $q_r^2 + p_r^2$. If e.g. there is a 1 : 1 frequency resonance (i.e. $\omega_1 = \omega_2$), then Ker(ζ) contains, in addition to $q_r^2 + p_r^2$, also $q_1 p_2 - q_2 p_1$ (and functions thereof). If there is a 1 : 2 resonance, i.e. $\omega_2 = 2\omega_1$, then Ker(ζ) contains also $2q_1p_1p_2 + (q_1^2 - p_1^2)q_2$.

Corollary 4. *If the quadratic part H_0 of the Hamiltonian has the form (34), and ω_r are not resonant, then the normal form $\widehat{H}$ of H is (Liouville) integrable. Indeed, $\widehat{H}$ is a function of the quantities $q_r^2 + p_r^2$ only, and these provide exactly ℓ constants of motion in involution for $\widehat{H}$.*

Remark 2. In the standard Birkhoff case (34) the homological operator ζ takes the form

$$\zeta = \sum \omega_r \left(q_r \frac{\partial}{\partial p_r} - p_r \frac{\partial}{\partial q_r} \right)$$

and this coincides with the operator introduced by Moser [246–248] (see also [17]).

For an abstract approach to NF of symmetric Hamiltonian systems, see also [235,347].

As already remarked for the case of generic dynamical problems, NF techniques are largely used in a great variety of physical problems. The classical application deals with celestial mechanics [174,247], with the natural extension to problems concerning the appearance of periodic or chaotic behavior (see, for example, [187,198,216,267,325] for some recent contributions). Results very similar to those holding for the Hamiltonian case are also available for *reversible* systems [64,213,294,295].

Although KAM theory is completely beyond the scope of this volume, let us quote some recent papers [111,112,233,244,245], which deal with this theory taking advantage from the use of NF techniques. We would also like to mention that recently the Gallavotti group introduced in this subject techniques derived from quantum field theory, and obtained in this way relevant results [42,152-154,156-158].

For applications of NF techniques to Hilbert spaces and quantum mechanics, we refer the reader to [170,171,258,281,284], and to [22,23] for an extension of the theory to abstract Banach spaces. Let us also mention [33] for an application to the Korteweg-de Vries equation, showing that the KdV Hamiltonian, and any Hamiltonian in the KdV hierarchy, admits a convergent Birkhoff normal form. Some different applications to specific physical problems will be cited in the next sections.

Exercise 7. Consider jointly the functions $H, K \in \mathcal{F}$ introduced in Exercise 5. Put them in normal form (i.e. with higher order terms of H, K in the kernel of the respective homological operator) up to terms in $\mathcal{F}_2$.

Exercise 8. Using the result of Exercise 7, put the same H and K in joint normal form, again up to terms in $\mathcal{F}_2$.

Exercise 9. Consider the quadratic Hamiltonian

$$K = \frac{1}{2}(p_1^2 + p_2^2) + \frac{1}{2}(q_1^2 + q_2^2) \ .$$

What are the Hamiltonians H which are in normal form and admit this K as a symmetry ?

Exercise 10. Consider the Hamiltonian

$$H = \frac{1}{2}(p_1^2 + p_2^2) + \frac{1}{2}(q_1^2 + q_2^2) + \frac{1}{4}(q_1^2 + q_2^2)^2 \ .$$

Is this in normal form ? Determine a nontrivial symmetry K for H, and consider the problem of writing jointly K and H in normal form, as in Exercise 7.

3. The Case of d-Dimensional Algebras of Symmetries

In the previous section, we have been studying only one symmetry, i.e. we had a pair (H, S) of commuting (under Poisson bracket) functions in $\mathcal{F}$. However, H could have more than one symmetry.

It is easy to check, by the Jacobi identity, that if S and R are both in $\mathcal{G}_H$, then $\{S, R\} \in \mathcal{G}_H$ as well, so that, as already mentioned, $\mathcal{G}_H$ is indeed a Lie algebra.[6] Obviously, we always have that $H \in \mathcal{G}_H$. The algebra $\mathcal{G}_H$ is not, in general, an abelian algebra. The only requirement on its structure is that H must be in the center of $\mathcal{G}_H$, by definition.

In the previous section we had that for an abelian two-dimensional algebra we can obtain a joint normal form. It is natural to ask if this result can be extended to more general situations.

Before discussing this point, we remark that the proofs of the result obtained in Sect. 3 of this chapter do quite closely follow the proofs of the corresponding results for Poincaré-Dulac NFs (given in Chap. IV). Such a similarity is not incidental, but depends on the fact that we only needed the formal properties of the Poisson bracket (the Lie-Poisson bracket in the Poincaré-Dulac case) and of the relevant homological operators.

Thus, it is clear that the same would apply to the results that we obtain in this section. Therefore, we just state the results that can be obtained in this way, referring for the proofs (and related remarks) to the corresponding ones for the Poincaré-Dulac case given in Chap. IV.

Theorem 6. *Let us consider a d-dimensional algebra $\mathcal{G} \subset \mathcal{F}$ of functions $F^{(1)}, ..., F^{(d)}$, with $F^{(i)} = B^{(i)} + \sum_{k=1}^{\infty} F_k^{(i)}$. Let us denote by β_i the homological operators $\beta_i(.) = \{B^{(i)}, .\}$, and assume that they are normal. Then, if the algebra $\mathcal{G}$ is nilpotent (in particular, abelian), then all the functions can be put in Joint Normal Form, namely transformed to functions $\widetilde{F}^{(i)} = B^{(i)} + \sum_{k=1}^{\infty} \widetilde{F}_k^{(i)}$ with*

$$\widetilde{F}_k^{(i)} \in \left[\bigcap_{j=1}^{d} \operatorname{Ker}(\beta_j) \right] \qquad (k > 0) \qquad \text{for each } i = 1, \dots, d \,.$$

Corollary 5. *In any nilpotent (in particular, abelian) subalgebra $\mathcal{G}_0$ of a generic algebra $\mathcal{G}$, all the functions spanning the subalgebra can be put in Joint Normal Form.*

A case of particular interest is that of H corresponding to an *integrable system*. In this case one can go from the (q,p) coordinates to action-angle coordinates (J, φ). In these new coordinates $\widehat{H} = \widehat{H}(J)$, so that the $J_1, ..., J_\ell$ are

6 Actually, multiplying a (analytic) symmetry by any (analytic) function of the symmetries themselves gives another (analytic) symmetry [and obviously the sum of two (analytic) symmetries is again a (analytic) symmetry], so one should more correctly say that we have a *module* of symmetries. Here, we are interested in the algebraic structure, and therefore we are considering only functionally independent (i.e. in particular pointwise linearly independent) symmetries. Notice also that some care is in order when considering this module structure in the Hamiltonian case, as the algebra of the Hamiltonian vector fields which are also Lie point symmetries does not have the structure of a module. Indeed, when multiplying a (Hamiltonian symmetry) vector field X_S by a constant of motion one does not obtain in general another Hamiltonian Lie point symmetry, but only a Lie point symmetry.

themselves constants of the motion, i.e. symmetries.[7] Moreover, the J commute among themselves as well as with H. Notice that now Theorem 6 does not give any restriction on $\widehat{H}$, as any function of J commutes with the J and is therefore in the relevant (intersection of) kernels. Note also that, however, in this case we have *linear* symmetries (in the (J, φ) coordinates), and – as H does not depend on the φ – we cannot have any more isolated stationary points; thus, our methods do not (directly) apply.

4. Perturbative Construction of Symmetries

In order to apply the results obtained so far, one should know (some of) the symmetries of the Hamiltonian under study. However, determining these can be a non-trivial problem, as it requires one to solve the PDEs $\{H, S\} = 0$.

We would like to point out that our results do *not* require a knowledge of the full symmetry algebra $\mathcal{G}_H$, but apply to any subset of $\mathcal{G}_H$. This means that we can apply them as soon as we know one symmetry, or only those symmetries which are "easy to find". Even so, the problem of determining the functions in $\mathcal{F}$ remains. It happens that if we look for them perturbatively (order by order), not only can we proceed to transformation into Joint (or Semisimple Joint, see below) NF at the same time, but also that the two procedures interact positively. By this, we mean that, on the one hand the determination of the symmetries helps in the normalization procedure, and on the other hand the transformation into NF helps in the determination of the symmetries. This point will be briefly discussed in this section (see also [98]). Clearly, our discussion is strongly related to that in Chap. IV for general dynamical systems.

We introduce the *approximate symmetries* for H. Let us first define the projectors $\Gamma_k : \mathcal{F} \to \mathcal{F}_k$ that associate an analytic function (or a formal power series) $F \in \mathcal{F}$ with its homogeneous part of degree k, i.e. F_k. Now we say that S is a symmetry of order k for H if $\Gamma_j(\{H, S\}) = 0$ for all the $j \leq k$.

One can check directly, by the Jacobi identity, that the symmetries of order k for a given H form a Lie algebra,[8] which we denote by $\mathcal{G}_H^{(k)}$. It is also obvious that $\mathcal{G}_H^{(k+1)} \subseteq \mathcal{G}_H^{(k)}$, so that considering the sequence of symmetries $\mathcal{G}_H^{(k)}$ for increasing k represents a filtration of Lie algebras [61–63,65].

The key observation is then that, if we consider normalization only up to order k, symmetries of order k are as good as exact ones. This means that we could restate all of our previous results with "NF up to order k" and "symmetries of order k" in the place of "NF" and "symmetries" tout court, and they would still hold with the same proof.

---

[7] The same holds, of course, for functions of the J's, in particular for the frequencies $\omega_i(J) = \dot{\varphi}_i = \partial H / \partial J_i$.

[8] Again, we also have a module structure.

Thus, suppose we have determined the approximate symmetry F up to order $(k-1)$, and have, accordingly, put H and this F in JNF[9] up to order $(k-1)$. We now have to go one order further to determine the symmetries, i.e. we have to solve (18) at order k. But we know that for all $m < k$ both H_m and F_m are in $\mathrm{Ker}(\zeta) \cap \mathrm{Ker}(\omega)$ (as H, F are in JNF up to order $k-1$), so that $\psi_k \in \mathrm{Ker}(\zeta) \cap \mathrm{Ker}(\omega)$ as well. We can now put H in NF up to order k, so that $\zeta(H_k) = 0$. Acting on both sides by ω and using again, as in Sect. 3 of this chapter, $[\omega, \zeta] = 0$, we get $\zeta(S_k) = 0$. This means that, rather than looking for solutions of (18) at order k in the whole space $\mathcal{F}_k$, we can just consider the smaller space $\mathcal{F}_k \cap \mathrm{Ker}(\zeta)$. If a solution S_k to this problem exists, we can put H and S in JNF up to order k, and start again the procedure for order $(k+1)$.

This shows that if we are interested in transforming our Hamiltonian in Birkhoff-Gustavson NF only up to some finite order k – as is always the case in real computations – we can proceed recursively to the determination of approximate symmetries and reduction into Joint (or Semisimple Joint, depending on the structure of the $\mathcal{G}_H^{(m)}$ which we determine) Normal Form, so that the results reported here can be algorithmically applied in practice.

In actual computations, moreover, one often considers a truncation $H^{(m)}$ of the relevant Hamiltonian H to some finite order m (and some stability result to ensure that the flows of H and of $H^{(m)}$ are similar enough over sufficiently short timescales). It is clear that in considering truncation of order m one has to consider approximate symmetries of the same order m as well, and in this case we also know that the general form of the NF expansion will obey the suitably modified versions of Theorems 1–6. Notice that, in particular, knowledge of approximate symmetries allows us to guarantee the vanishing of terms which would be, in general, present in the NF expansion, and could therefore guarantee stability over longer timescales than in the absence of the symmetries. Such a discussion, however, goes beyond the limits of this work and will not be pursued here. Let us also remark that the knowledge of even only the first terms of the NF can be useful for physical interpretation, e.g. for determining the local stability of the solution [293,311]. For an important example, we refer to [37], where the theory of the betatronic motion of a particle in a circular accelerator is analyzed with great accuracy. Stability problems appear in many other fields, e.g. in celestial and satellite mechanics, see for example [296]. For an interesting example of the relationship between truncated NF and symmetries, see [303], where the classical Hénon-Heiles Hamiltonian is considered. Other applications deal with the description of vibrational transition modes of coupled oscillators [1].

Exercise 11. Determine at least one approximate symmetry (up to terms in $\mathcal{F}_4$) of

[9] The argument would go along the same lines for JNF of more than two functions. Here we concentrate on the JNF of two functions for the sake of simplicity.

$$H = \frac{1}{2}(p_1^2 + p_2^2) + \frac{1}{2}(q_1^2 + q_2^2) + [q_1^4 + 2q_1^2q_2^2 + q_2^4 + q_1^6 + 3q_1^4q_2^2 + 2q_1^2q_2^4 + q_2^6] \ .$$

Exercise 12. Consider again the Hamiltonian K introduced in Exercise 5. Determine approximate symmetries of this, and restrict consequently the normal form expansion for it using the approach outlined in this section.

5. The Non-normal Case

In Sects. 2 and 3 of this chapter, we have assumed the normality of the relevant homological operators. We want now to discuss how our results are modified when we relax this assumption.

As discussed in the next section, the relevant matrix for the normality of ζ is $A = JC$, where J is the standard $(2\ell \times 2\ell)$ symplectic matrix,

$$J = \begin{pmatrix} 0 & I \\ -I & 0 \end{pmatrix} \tag{35}$$

and C is the real matrix, which we can always take as being symmetric, defining H_0 by

$$H_0 = \frac{1}{2}\, C_{jk}\, x_j\, x_k \ . \tag{36}$$

We then recall a standard result of linear algebra [183], already stated in Chap. IV, which ensures that

Lemma 5 (S+N decomposition). *A real matrix A can be uniquely decomposed as $A = A_s + A_n$ where A_s is semisimple, A_n is nilpotent, and $[A_s, A_n] = 0$.*

By an easy computation (Chap. IV) we also have

Lemma 6. *If $[A, B] = 0$ and $A = A_s + A_n$, $B = B_s + B_n$ give the unique S+N decompositions for A and B, then $[A_s, B] = [A, B_s] = 0$.*

We can then apply these results to the matrices which are relevant for the homological operators, i.e. to $A = JC$ for the operator ζ associated with the Hamiltonian (36), and more in general (i.e. as in Sect. 3) with the matrices $A^{(i)} = JC^{(i)}$ for the operators β_i associated with general quadratic functions, $B^{(i)} = \frac{1}{2}C^{(i)}_{jk}x_jx_k$. Thus, we write the unique $S+N$ decompositions of these matrices as

$$A^{(i)} = A^{(i)}_s + A^{(i)}_n \ . \tag{37}$$

Correspondingly, by defining

$$C^{(i)}_S \ = \ J^+A^{(i)}_s \quad ; \quad C^{(i)}_N \ = \ J^+A^{(i)}_n \ , \tag{38}$$

we can introduce the "S part" and the "N part" of the quadratic functions $B^{(i)}$ according to

$$B_S^{(i)} = \frac{1}{2} \left[C_S^{(i)} \right]_{jk} x_j \, x_k \quad ; \tag{39}$$

with

$$B^{(i)} = B_S^{(i)} + B_N^{(i)} . \tag{40}$$

This is turn yields a unique decomposition of the homological operators β_i associated with $B^{(i)}$ as

$$\beta_i = \beta_{i,s} + \beta_{i,n} \quad ; \quad \beta_{i,s} = \{B_S^{(i)}, \cdot\} \ , \ \beta_{i,n} = \{B_N^{(i)}, \cdot\} \ , \tag{41}$$

and similarly for the operator ζ associated with H_0, i.e. $\zeta = \zeta_s + \zeta_n \ ; \ \zeta_s = \{H_S, \cdot\} \ , \ \zeta_n = \{H_N, \cdot\}$. We call this decomposition the *S+N induced decomposition for homological operators.*

With this construction and notation, we have the following definition and theorem.

Definition 3. *Let the algebra $\mathcal{G}$ be spanned by the functions $F^{(1)}, ..., F^{(d)} \in \mathcal{F}$, and let $B^{(i)}$ be the quadratic part of $F^{(i)}$: $F^{(i)} = B^{(i)} + F_R^{(i)}$. Denote by β_i the homological operator associated with $B^{(i)}$, and let $\beta_i = \beta_{i,s} + \beta_{i,n}$ be its S+N induced decomposition. If all the $F^{(i)}$ are in*

$$\mathcal{K} = \bigcap_{i=1}^{d} \mathrm{Ker}\,(\beta_{i,s}) \ ,$$

we say that $\mathcal{G}$ is in Semisimple Joint Normal Form. If, for each fixed i, one has

$$F_R^{(i)} \in \Big(\bigcap_{j \neq i} \mathrm{Ker}(\beta_{j,s}) \Big) \cap \mathrm{Ker}(\beta_i^+)$$

and for all other $j \neq i$, $F^{(j)}$ are in $\mathcal{K}$, we say that $\mathcal{G}$ is in i-Joint Normal Form.

Theorem 7. *Let us consider a d-dimensional algebra $\mathcal{G}$ of functions in $\mathcal{F}$, spanned by $F^{(1)}, ..., F^{(d)}$ with $F^{(i)} = B^{(i)} + \sum_{k=1}^{\infty} F_k^{(i)}$. Let us denote by β_i the homological operators $\beta_i = \{B^{(i)}, .\}$, let $\beta_i = \beta_{i,s} + \beta_{i,n}$ be their S+N induced decomposition, and assume that the "S part" $\beta_{i,s}$ are normal. Then, if the algebra $\mathcal{G}$ is* nilpotent *(in particular, abelian), then all the functions $F^{(i)}$ can be put in Semisimple Joint Normal Form by a series of Birkhoff transformations, or also, for each fixed a, in a-Joint Normal Form.*

Corollary 6. *In any nilpotent (in particular, abelian) subalgebra $\mathcal{G}_0$ of a generic algebra $\mathcal{G}$, all the functions spanning the subalgebra can be put in Semisimple Joint Normal Form.*

Proof. See the proof of Theorems 5 and 6 in Chap. IV. ⊙

Example 1. Let us consider the 3-dimensional nilpotent algebra spanned by $F^{(1)}, F^{(2)}, H$ satisfying

$$\{F^{(1)}, H\} = 0, \qquad \{F^{(2)}, H\} = 0, \qquad \{F^{(1)}, F^{(2)}\} = H \ . \tag{42}$$

Let β_1, β_2, ζ be the associated homological operators, and let the "S part" $\beta_{a,s}$ of β_a be a normal operator. Then Theorem 7 gives

$$H_R \to \widehat{H}_R \in \mathrm{Ker}(\zeta^+) \cap \mathrm{Ker}(\beta_{1,s}) \cap \mathrm{Ker}(\beta_{2,s})$$
$$F^{(1)},\ F^{(2)} \to \widehat{F}^{(1)},\ \widehat{F}^{(2)} \in \mathrm{Ker}(\beta_{1,s}) \cap \mathrm{Ker}(\beta_{2,s})\ .$$

Notice that from the commutator of the quadratic parts, i.e. $\{F_0^{(1)}, F_0^{(2)}\} = H_0$, one easily obtains that the "S-part" H_S of H_0 is zero, and therefore no information can be derived in this case from the other result $\widehat{H}, \widehat{F}^{(1)}, \widehat{F}^{(2)} \in \mathrm{Ker}(\zeta_s)$. For instance, the following quadratic functions, in $\ell = 3$ degrees of freedom,

$$F_0^{(1)} = q_2 p_1 \quad , \quad F_0^{(2)} = q_3 p_2 \quad , \quad H_0 = q_3 p_1$$

satisfy (42), but with this choice, $\beta_{1,s} = \beta_{2,s} = \zeta_s = 0$, and no information can be obtained from the Semisimple JNF. Modifying, e.g., only the choice of $F_0^{(1)}$ in this way

$$F_0^{(1)} = q_2 p_1 + q_1 p_1 + q_2 p_2 + q_3 p_3$$

the commutation rules are unchanged, but now $\mathrm{Ker}(\beta_{1,s})$ contains all monomials of the form $q_i p_j$ and $\mathrm{Ker}(\zeta^+)$ contains functions of $q_1, q_2, p_2, p_3, q_i p_i$. Therefore, in any 3-dimensional nilpotent algebra satisfying (42) and having as quadratic parts the above terms, all higher-order terms $\widehat{H}_R$ of the Hamiltonian must depend, once in NF, only on the quantities $q_1 p_2,\ q_1 p_3,\ q_2 p_2,\ q_2 p_3,\ q_i p_i$, and $F^{(1)},\ F^{(2)}$ only on the monomials $q_i p_j$.

In the case of algebras of functions which are not nilpotent (these are necessarily non-abelian), the corresponding theorems for DS given in Chap. IV can be easily translated into the Hamiltonian language, and can also have some interest in Hamiltonian problems. Indeed they deal with algebras which admit a nontrivial center and, as already noted, the algebra $\mathcal{G}_H$ of the symmetries of H always contains, by definition, a center $C(\mathcal{G}_H)$ that includes the Hamiltonian itself. For instance, the case covered by Theorem 10 of Chap. IV can be of interest: Hamiltonian which admits a solvable n-dimensional algebra of symmetries is integrable by quadratures, under suitable additional assumptions, according to a theorem by Kozlov and Kolesnikov [17]. Instead of repeating the general statements in Hamiltonian language, let us restrict ourselves, for the sake of simplicity, to the cases of lower dimensions. We get for instance from Theorems 8 or 10 of Chap. IV:

Corollary 7. *Let $F^{(1)}, F^{(2)}$ span a 2−dimensional algebra commuting with H:*

$$\{F^{(1)}, F^{(2)}\} = F^{(2)} \quad , \quad \{F^{(a)}, H\} = 0 \qquad (a = 1, 2)\ .$$

Then there is a formal Birkhoff transformation which takes $H, F^{(a)}$ into the form

$$H \to \widehat{H} \in \mathrm{Ker}(\zeta^+) \cap \mathrm{Ker}(\beta_{1,s}) \cap \mathrm{Ker}(\beta_{2,s})$$
$$F^{(1)}, F^{(2)} \to \widetilde{F}^{(1)}, \widetilde{F}^{(2)} \in \mathrm{Ker}(\zeta_s)$$

where, as usual, ζ_s is the semisimple part of ζ.

Remark 3. If, e.g., $F_0^{(1)}$ has a nonzero "S-part", i.e. $\beta_{1,s} \neq 0$, then also the quadratic "S-part" $F_S^{(1)}$ of $F_0^{(1)}$ is a symmetry of $\widehat{H}$ in NF.

Example 2. In $\ell = 2$ degrees of freedom, let

$$F_0^{(1)} = q_1 p_1 \quad , \quad F_0^{(2)} = q_2 p_1$$

and

$$H = H_0 + H_R \quad \text{with} \quad H_0 = q_1 p_1 + q_2 p_2 \quad , \quad H_R = H_R(q_1 p_1 + q_2 p_2,\ q_2) \ .$$

It is easy to verify that all conditions of Corollary 7 are met by this choice. In this example, $\zeta = \zeta_s$, $\beta_1 = \beta_{1,s}$, $\beta_{2,s} = 0$. Now, Corollary 7 implies that in NF, $\widehat{H}$ (and $\widetilde{F}^{(1)}, \widetilde{F}^{(2)}$ as well) are in $\mathrm{Ker}(\zeta)$. In addition, Theorem 2 guarantees that the quadratic functions $F_0^{(1)}$ and $F_0^{(2)}$ are symmetries of $\widehat{H}$. These facts imply that the NF $\widehat{H}$ must be a function of $q_1 p_1 + q_2 p_2 = H_0$ only.

Finally, for the case of solvable algebras, one can get, from Theorem 9 of Chap. IV in dimension $n = 2$ (notice that 2-dimensional algebras are either abelian or solvable) the following:

Corollary 8. *Let $F^{(1)}, F^{(2)}$ span a (solvable) 2-dimensional algebra:*

$$\{F^{(1)}, F^{(2)}\} = F^{(2)}$$

(where either $F^{(1)}$ or $F^{(2)}$ can be assumed to be the Hamiltonian). Then there is a formal Birkhoff transformation which takes the higher-order terms $F_R^{(1)}$, $F_R^{(2)}$ in the form

$$F_R^{(1)} \to \widetilde{F}_R^{(1)} \in \mathrm{Ker}(\beta_2) \quad \text{and} \quad F_R^{(2)} \to \widehat{F}_R^{(2)} \in \mathrm{Ker}(\beta_2^+) \ .$$

Exercise 13. Let $F^{(1)}, F^{(2)}$ span a solvable algebra as in Corollary 8, and assume that their quadratic parts are

$$F_0^{(1)} = q_1 p_1 \qquad F_0^{(2)} = q_2 p_1 \ .$$

Show that the higher-order terms $F_R^{(1)}$ of $F^{(1)}$, once in NF, must depend only on $q_2, p_1, q_i p_i$. What about $F_R^{(2)}$?

6. The Normality of the Homological Operator

The normality of the homological operator $\zeta = \{H_0, .\}$ plays a crucial role in the normalization procedure, and more in general in NF theory. We are going to analyze it in this section.

With the notation introduced in the previous section, it will turn out that:

Lemma 7. *Let ζ be the homological operator associated with a quadratic Hamiltonian $H_0 = (1/2)C_{ij}x_ix_j$ (in which $C = C^T$), i.e. $\zeta = \{H_0, .\}$. Then, $[\zeta, \zeta^+] = 0$ if and only if $[A, A^+] = 0$, where $A = JC$.*

Proof. If $B^{(i)} = (1/2)(x, C_i x)$ are quadratic functions [where clearly $(\cdot,\cdot)$ stands for the usual scalar product in $R^{2\ell}$], and $\beta^{(i)}$ are the associated homological operators,

$$\beta^{(i)} = \{B^{(i)}, \cdot\} ,$$

then one easily obtains

$$[\beta^{(i)}, \beta^{(j)}] = \beta^{(k)} \qquad \text{if} \qquad \{B^{(i)}, B^{(j)}\} = B^{(k)}$$

and, putting $A_i = JC_i$,

$$\{B^{(i)}, B^{(j)}\} = \frac{1}{2}(x, (C_iJC_j - C_jJC_i)x\ ) = -\frac{1}{2}(x, J[A_i, A_j]\ x) \ . \tag{43}$$

The result then follows choosing $B^{(1)} = H_0 = (1/2)(x, Cx)$ and $B^{(2)} = H_0'$ where H_0' is the "adjoint" Hamiltonian defined by

$$H_0' = \frac{1}{2}(x,\ JCJ\ x) = \frac{1}{2}(x,\ C'\ x) \tag{44}$$

in such a way that $\dot{x} = J\nabla H_0' = A^+x$ and $A_1 = A = JC$, $A_2 = JC' = A^+$, and observing that the adjoint operator ζ^+ is precisely the operator associated with the Hamiltonian H_0'. ⊙

Having proved that the normality of ζ reduces to the normality of JC, it remains to study the conditions on C that ensure this to be the case. For an explicit matrix C, of course, we only have to consider the product JC and check explicitly if this is normal. We can write C in block form

$$C = \begin{pmatrix} a & b \\ b^+ & c \end{pmatrix} \tag{45}$$

where a, b, c are real $(\ell \times \ell)$ matrices, and $a = a^+$, $c = c^+$. The condition $[A, A^+] = 0$ is then

$$\begin{cases} [b^+, b] = a^2 - c^2 \\ (ab + b^+a) + (cb^+ + bc) = 0 \end{cases} . \tag{46}$$

For a general discussion, the above formula is not convenient, and one should better consider the *normal forms for quadratic Hamiltonians* [13,17]. E.g., in the case of $\ell = 1$ degree of freedom, ζ is semisimple if and only if H has no zero eigenvalue, and in this case H_0 can be canonically put in one of the forms

$$H_0 = \pm\frac{c^2}{2}(p^2 \pm q^2)$$

that correspond to normal ζ. The Hamiltonian $H_0 = p^2/2$ gives an obvious example of non-normal (nor diagonalizable) A.

For an explicit non-trivial and physically relevant example of a situation of this type, we refer the reader to [122], where the case of a "magnetic bottle", made up of a magnetic dipole field with a superimposed octupole contribution, has been examined in full detail.

If the degrees of freedom are $\ell > 1$, a class of Hamiltonians giving normal ζ is the following

$$H_0 = \pm\frac{1}{2}\Big(\sum_r c_r^2(p_r^2 + q_r^2) + \sum_s c_s'^2(p_s^2 - q_s^2)\Big)$$

extending the standard Birkhoff class $H_0 = \frac{1}{2}\sum\omega_r(p_r^2+q_r^2)$. All quadratic Hamiltonians with purely imaginary and distinct eigenvalues $\pm i\omega_r$ can in fact be reduced to the Birkhoff form ([17], p.111). Different examples can be given: e.g. in $\ell = 2$ the following Hamiltonian

$$H_0 = q_1(p_1 - p_2) \ + \ q_2(p_1 + p_2)$$

with eigenvalues $\pm 1 \pm i$, corresponds to a normal ζ. Also, all H_0 with algebraically simple eigenvalues can be canonically transformed into Hamiltonians giving normal ζ [17,340]. This follows from a theorem of Williamson [340], that we quote from chapter 7.2.3 of [17]:

Theorem 8 (Williamson). *The quadratic Hamiltonian H_0 can be reduced by a symplectic change of variables to a sum of partial Hamiltonians, functions of different sets of canonically conjugate variables. The matrix C of the corresponding Hamiltonian system is thereby reduced to block form. Each partial Hamiltonian corresponds to either a real pair, a purely imaginary pair, a quadruple of eigenvalues, or to a zero eigenvalue. The partial Hamiltonians are uniquely determined, up to a sign, by the Jordan blocks of the matrix JC.*

Thus, we can go to the coordinates corresponding to the splitting of H_0 in partial Hamiltonians provided by the theorem above. We will call (q_i, p_i) the pairs of canonical coordinates, and have these ordered in groups corresponding to zero eigenvalues, type I eigenvalues, etc., denoting the corresponding sets of indices as α_0, α_1, α_2, α_3.

In the generic case of simple eigenvalues, one can then write H_0 as [17]

$$H_0 = \sum_{i\in\alpha_1} a_i(p_i q_i) + \sum_{i\in\alpha_2} a_i(p_i^2 + q_i^2) + \sum_{i,j\in\alpha_3} [-a_i(p_i q_i + p_j q_j) + b_i(p_i q_j - p_j q_i)] \tag{47}$$

(in the α_3 sum, the i and j must correspond to the same quadruplet, say $j = i+1$).

We can then consider the normality of A block by block. For types 0, I, II it is clear that we get normal matrices. For blocks of type III, the normality follows from a simple computation.

It should be stressed, however, that the presence of symmetry can alter the generic properties of H_0, so that in the present setting we cannot claim that the case of normal ζ is generic.

Examples of Hamiltonians with non-normal (nor semisimple) A are the following (with $\ell = 2$)

$$H_0 = -(q_1p_1 + q_2p_2) \; + \; p_1q_2$$

having eigenvalues ± 1 with double algebraic multiplicity, or

$$H_0 = \frac{1}{2}(p_1^2 + p_2^2) + q_1p_2 - q_2p_1$$

with algebraically double eigenvalues $\pm i$.

VI. Convergence of the Normalizing Transformations

As is well known, the normalizing transformation (NT) – that is, the series of Poincaré transformations, and of Birkhoff transformations as well, needed to bring a given VF $f(x)$ (or DS) and respectively the given Hamiltonian $H(x)$ into NF – is in general only formal, as we are not sure that the radius of convergence of the series is greater than zero. It would definitely be of interest to characterize convergence or divergence of the NT in terms of the DS or of the Hamiltonian. We shall see that the symmetry properties of the system can play some role in this problem. The first four Sections of this chapter are devoted to the general case of DS, whereas in Section 5 we will consider the special case of Hamiltonian problems.

1. Conditions Ensuring Convergence

A well known sufficient condition that ensures convergence of the NT is given by the following Poincaré criterion (see [12,16] for more details, and also [108,137] for some recent contributions to the analysis of the domains of analyticity). Unless otherwise stated, in the present chapter, the matrix $A = (Df)(0)$ is assumed to be semisimple.

Theorem 1 (Poincaré). *If the eigenvalues $\lambda_1, \ldots, \lambda_n$ of the matrix $A = (Df)(0)$ of the analytic VF f belong to the Poincaré domain, i.e. if the convex hull of the n points $\lambda_1, \ldots, \lambda_n$ in the complex plane does not contain zero, then the NT is convergent and the NF is analytic.*

There are also two basic and very general conditions, given by Bruno, and called Condition A and Condition ω (see [66,67,69] for details), which ensure that a vector function $f = Ax + F$ can be put in NF by a converging transformation. For convenience, we state here Condition A in its simplest (and quite restrictive) form. Let us assume for the moment that there is a straight line through the origin in the complex plane that contains all the eigenvalues λ_i of A, and that there are eigenvalues lying on both parts of this line with respect to the origin. In other words, all the eigenvalues are proportional through real factors, $\lambda_i = c_i \lambda_0$ where λ_0 is some complex number and $c_i \in \mathbf{R}$, the signs of the c_i's being not all equal (otherwise the λ_i would belong to a Poincaré domain and convergence would be guaranteed by Theorem 1 above). Note that when we consider, as we do, real matrices, this condition can be verified only if the eigenvalues are either all real or all pure imaginary. Then Condition A reads

Condition A. *A normal form $\widehat{f}$ is said to satisfy Condition A if $\widehat{f}$ has the form*

$$\widehat{f} = Ax + \alpha(x)Ax \tag{1}$$

where $\alpha(x)$ is some scalar-valued power series.

Actually, depending on the position of the eigenvalues of A in the complex plane, there are different versions of Condition A that allow for the presence of other terms in the r.h.s. of (1), and the above statement may be modified accordingly [66,67,69]. Let us note that in all the applications below, where in particular only linear NF will be ultimately concerned, the above formulation of Condition A is enough to cover all the cases to be considered.

The other condition is:

Condition ω. *Let $\omega_k = \min |(Q,\Lambda) - \lambda_j|$ for all $j = 1,\dots,n$ and all n-tuples of nonnegative integers q_i such that $1 < \sum_{i=1}^{n} q_i < 2^k$ and $(Q,\Lambda) = \sum_i q_i\lambda_i \neq \lambda_j$. Then Condition ω is satisfied if*

$$\sum_{k=1}^{\infty} 2^{-k} \ln\left(\omega_k^{-1}\right) < \infty \, . \tag{2}$$

This is a very weak condition, devised to control the appearance of small divisors in the series of NT, and generalizes the Siegel-type condition (see [12,66–69]):

$$|(Q,\Lambda) - \lambda_j| > \varepsilon \left(\sum_{i=1}^{n} |q_i|\right)^{-\nu}$$

for some $\varepsilon, \nu > 0$, or the much simpler condition $|(Q,\Lambda) - \lambda_j| > \varepsilon > 0$, for all n-tuple q_i such that $(Q,\Lambda) \neq \lambda_j$.

With these two conditions, one then has [66,67,69]:

Theorem 1′ (Bruno). *If $A = (Df)(0)$ satisfies Condition ω, and f can be transformed, via P transformations, to a NF $\widehat{f}$ which satisfies Condition A, then there is a convergent NT for f.*

Clearly, here and in the following, "convergence" stands for "convergence in some open neighborhood of the stationary point $x_0 = 0$". We will also assume that condition ω is always satisfied (see, however, [139]), and concentrate on Condition A; i.e. we take for granted from now on that

Assumption B. *The matrix $A = (Df)(0)$ satisfies Condition ω.*

It is easy to prove that

Lemma 1. *If f (not necessarily in NF) is of the form* (1), *then a VF of the form $g = \mu(x)\ Bx$ is a symmetry for f, i.e. $g \in \mathcal{G}_f$, if the following conditions are satisfied:* i) $[A,\ B] = 0$, ii) $\mu(x) \in \mathcal{I}_A^*$, iii) $\alpha(x) \in \mathcal{I}_B^*$.

Indeed, if $f = \big(1 + \alpha(x)\big)\ Ax$ and $g = \mu(x)Bx$, we have

$$\{f,g\} = (1+\alpha)\mu[A,\ B]x + (1+\alpha)X_A(\mu)Bx - \mu X_B(\alpha)Ax = 0 \, .$$

Lemma 2. *Let a NF $\widehat{f}$ satisfy Condition A: then $\alpha(x) \in \mathcal{I}_A^*$.*

Indeed, if f is in NF and satisfies Condition A, then $[X_A,\ X_f] = 0$ implies that $X_A(\alpha) = 0$, so that $\alpha(x) \in \mathcal{I}_A$. Actually, this result is a special case of Lemma 13 of Chap. IV (being obviously $A \in \mathcal{L}_A$). As already remarked, if f has the form (1), then trivially $\mathcal{I}_f = \mathcal{I}_A$ and $\mathcal{I}_f^* = \mathcal{I}_A^*$. Notice that, even if f in (1) is in NF, i.e. $X_A(\alpha) = 0$, then $[A,\ B] = 0$ ensures that $X_A(X_B(\alpha)) = 0$, but not necessarily $X_B(\alpha) = 0$. Indeed, it suffices to consider $B = I$ to become convinced that iii) must be explicitly assumed in Lemma 1.

Lemma 3. *Let the DS* (1) *be in NF and satisfy Condition A; let g be a symmetry of the DS. Then the constant of motion $\alpha(x)$ also satisfies $g \cdot \nabla\alpha = 0$, i.e. $\alpha \in \mathcal{I}_g^* \cap \mathcal{I}_f^*$.*

Indeed, from the symmetry condition, one gets $\{f, g\} = \alpha\mathcal{A}(g) - (g \cdot \nabla\alpha)Ax = 0$, and the result follows from the property $g \in \mathrm{Ker}(\mathcal{A})$ shown in Theorem 1 of Chap. IV.

Even if bifurcation theory is not a specific topic of this book (except some remarks in Chap. II), let us mention a useful physical application (with some recent developments [90]) in bifurcation theory of Condition A.

Consider a DS, depending also on some real "control parameters" $\mu \in \mathbf{R}^p$

$$\dot{x} = A(\mu)x + F(\mu, x)$$

and assume that, for some "critical" value $\mu = \mu_0$ of the parameters, the matrix $A(\mu_0)$ is semisimple and its eigenvalues $\lambda_i(\mu_0)$ satisfy a resonance relation. Then, introducing the NF of this DS and considering the parameters μ as independent variables, the idea is to impose that Condition A is satisfied: This leads to some prescriptions which – under suitable hypotheses – can be fulfilled thanks to the presence of the parameters μ. Then, the problem in NF can be easily solved, and, coming back to the initial coordinates (this in fact can be performed by a convergent transformation), a bifurcating solution is obtained, in correspondence to the resonant eigenvalues $\lambda_i(\mu_0)$. Using this approach via NF, not only standard Hopf and stationary bifurcations are recovered, but also some quite different and general situations can be dealt with [90]. A typical and physically relevant example describes coupled oscillators with resonant unperturbed frequencies: we can state e.g. the following result (which we quote here without proof, see [90]).

Theorem 2. *Consider a 4-dimensional DS depending on 3 real parameters $\mu \in \mathbf{R}^3$, and assume that for some $\mu = \mu_0$ the eigenvalues of the matrix $A(\mu_0)$ are $\lambda = \pm i\omega_0,\ \pm im\omega_0$ (with $\omega_0 > 0$, $m = 2, 3, \ldots$). After complexification of the space, let $A^C(\mu)$ be the matrix $A(\mu)$ written in the basis in which $A^C(\mu_0)$ is diagonal. Putting now $a_k^{(i)} = \left.\dfrac{\partial A_{ii}^C(\mu)}{\partial \mu_k}\right|_{\mu=\mu_0}$ $(i = 1, \ldots, 4\ ;\ k = 1, 2, 3)$, assume that the 4×4 matrix*

$$D = \begin{pmatrix} a_1^{(1)} & a_2^{(1)} & a_3^{(1)} & 1 \\ a_1^{(2)} & \ldots & & -1 \\ \ldots & \ldots & & m \\ a_1^{(4)} & \ldots & & -m \end{pmatrix}$$

is not singular, i.e. $\det(D) \neq 0$; *then there is a multiple-periodic bifurcating solution preserving the frequency resonance* $1 : m$.

The method can be suitably extended to cover cases of several oscillators and also of degenerate eigenvalues and frequences in the presence of symmetry [90].

We may also illustrate, by means of an example in the same context, the crucial role played by Condition A in convergence problems. Consider the following problem, which is essentially an adjustment (in view of the present discussion) of an example given, with quite different intent, in [185]: with $\mathbf{x} \equiv (x, y, z) \in \mathbf{R}^3$ and $\mu \in \mathbf{R}^2$, let

$$\dot{\mathbf{x}} = A(\mu)\mathbf{x} + \begin{pmatrix} -(r^2+3z^2)x \\ -(r^2+3z^2)y \\ (3r^2+z^2)z \end{pmatrix} \qquad \text{where} \qquad A(\mu) = \begin{pmatrix} \mu_1+\mu_2 & 1 & 0 \\ -1 & \mu_1 & 0 \\ 0 & 0 & -\mu_1 \end{pmatrix}$$

and $r^2 = x^2 + y^2$. It is not difficult to see (cf. [185]) that, if $\mu_2 = 0$, this system admits a family of heteroclinic orbits connecting the origin to the circle $r^2 = \mu_1$ and another family of heteroclinic orbits, living on the manifold $r^2 + z^2 = \mu_1$, connecting the circle $r^2 = \mu_1$ to the point $P \equiv (0, 0, \mu_1)$. When $\mu_2 \neq 0$, this heteroclinic structure breaks down, and an application of the Melnikov theory [173,185] shows the occurrence of transversal intersections of stable and unstable manifolds, with the consequent appearance of the chaotic behavior described by the classical Birkhoff-Smale horseshoe-like structure (actually, we need here a simple 3-dimensional version of the standard Melnikov theorem: see, for example, [185]). On the other hand, the NF of the above DS, evaluated at the critical value $\mu_0 = 0$ of the parameters, exhibits a perfectly regular (i.e., non-chaotic) behavior: indeed, the NF must possess the linear symmetry generated by $A(0)$, which implies that the NF is symmetric under rotations around the z axis. Then, the NF is essentially a 2-dimensional problem and therefore no chaos is allowed, as a consequence of the classical Poincaré-Bendixson theorem. This clearly implies that the NT cannot be convergent.

If one now uses the parameters μ to impose Condition A on this NF, it is easily seen that this condition is satisfied only along the circle defined by

$$\mu_1 = 3(x^2 + y^2) + z^2 \qquad\qquad \mu_2 = 4(z - x^2 - y^2)$$

(in the new coordinates) where in fact a (Hopf-type) periodic solution occurs. Then, in conclusion, we are here in the presence of a regular solution (where Condition A is indeed satisfied), which is completely surrounded by chaotic solutions.

This clearly confirms the relevance of Condition A in these arguments. In the next sections, we shall show that, even if the given DS does not satisfy Condition A, one can equally get some useful results thanks now to the presence of some symmetry properties of the problem.

2. Normalizing Transformations in the Presence of Symmetries

Let us restate some of our above Lemmata in a more convenient way (cf. Lemma 13 in Chap. IV, and resp. Corollaries 1 and 2 in Chap. IV, see also [66,67,330]):

Lemma 4. *Given the matrix A, and choosing a basis M_j in the space of the matrices M commuting with A, the set of the terms $\widehat{F}(x) \in \mathrm{Ker}(\mathcal{A})$ is given by a power series of the form*

$$\widehat{F} = \sum_j \mu_j(x) M_j x \, , \tag{3}$$

where $\mu_j(x) \in \mathcal{I}_A^*$ are constants of motion (expressed by a power series or also possibly by rational functions, see [330] for a detailed statement) of the linear system $\dot{x} = Ax$.

Lemma 5. *If the original DS $\dot{x} = f(x) = Ax + F(x)$ admits a linear symmetry $g_B = Bx$, then there is a NF $\widehat{f}$ which admits the same symmetry g_B, and this NF actually satisfies $\widehat{f} \in \mathrm{Ker}(\mathcal{A}) \cap \mathrm{Ker}(\mathcal{B})$. If f admits a symmetry $g = Bx + G(x)$, where also B is semisimple, then the linear part Bx of g is a symmetry of the DS once in JNF.*

Remark 1. The linear VF Bx is a symmetry for the DS $\dot{x} = Ax + \widehat{F}$ if and only if, with the notations of (3), $[B, A] = [B, M_j] = 0$ and $(Bx) \cdot \nabla \mu_j = 0, \forall j$. Indeed, Bx is a symmetry for f if and only if $f \in \mathrm{Ker}(\mathcal{B})$, and the result follows from Lemma 13 of Chap. IV.

As we did for the matrix A, we will assume that B satisfies Condition ω, and also – for the sake of simplicity, and unless otherwise stated – that B is semisimple. A simple but useful result, based on the possible introduction of a convergent NT for the symmetry field g, is then:

Theorem 3. *Let the DS $\dot{x} = f = Ax + F$ admit an analytic symmetry $g = Bx + G(x)$ with B not proportional to A. Assume that either:* i) *the eigenvalues of B belong to the Poincaré domain, or* ii) *there is a coordinate transformation taking g into a NF $\widehat{g}$ satisfying Condition A (e.g., with a linear $\widehat{g} = Bx$). Then, there is a convergent coordinate transformation which takes (not only g into NF but also) the DS into the following form:*

$$f \to Ax + \widetilde{F}(x) \quad \text{where} \quad \widetilde{F} \in \mathrm{Ker}(\mathcal{B}) \, ,$$

i.e. where $\widetilde{F}$ is in NF with respect to B (not necessarily to A).

According to Theorem 1 of Chap. IV, another (possibly formal) coordinate transformation can put $\widetilde{F}$ into normal form also with respect to A (JNF), but it is not granted that this other transformation is convergent.

The above theorem can be useful, e.g., in the special case that $\mathrm{Ker}(\mathcal{B}) = \{0\}$ (e.g. if $B = I$), and therefore in the problem of the linearization of the DS. In fact, we can now restate and complete some results given above (Chap. IV) in the following form [24,150]:

Corollary 1. *A DS* $\dot{x} = f = Ax + F$ *can be linearized if and only if it admits a (possibly formal) symmetry* $g = Bx + G$ *such that* $B = Dg(0) = I$, *and the NT that linearizes* f *is convergent if and only if there is an analytic symmetry of this type. More in general, if* g *is analytic and* $\mathrm{Ker}(\mathcal{B}) = \{0\}$, *then the NT which linearizes* f *is convergent.*

Notice that, if g is analytic, the convergence of the NT is guaranteed by the Poincaré Theorem in the case $B = I$, and by the Bruno Theorem (in particular, by Condition A applied to g) in the other general case $\mathrm{Ker}(\mathcal{B}) = \{0\}$.

Even if quite simple and well known, a class of examples which can be usefully reconsidered from this point of view is provided by the "holomorphic" 2-dimensional dynamical systems, i.e. by DS of the form

$$\dot{x} = u(x, y) \qquad\qquad \dot{y} = v(x, y)$$

where, putting $z = x + iy$, the function $f(z) = u + iv$ is a holomorphic function of the complex variable z. It can be easily verified that these DS admit the non-trivial analytic symmetry

$$g \cdot \nabla = v(x, y)\frac{\partial}{\partial x} - u(x, y)\frac{\partial}{\partial y} \ .$$

Now, writing $f(z) = \alpha z + F(z)$, if $\mathrm{Re}\ \alpha \neq 0$ then $\mathrm{Ker}(\mathcal{A}) = \{0\}$ and the DS can be linearized by means of a convergent transformation (thanks to the Poincaré criterion). Consider instead the more interesting case $\mathrm{Re}\ \alpha = 0$ (with $\alpha \neq 0$): then the eigenvalues of A are complex conjugate and $\mathrm{Ker}(\mathcal{A}) \neq \{0\}$, but now the symmetry is such that $B = I$ (or proportional to I), and again – according to the above corollary – the DS can be put into linear form by a convergent transformation. Also, notice that in the case $\mathrm{Re}\ \alpha = 0$, the holomorphic DS are of some physical interest because they describe the "isochronous centers" [5,292] i.e. planar systems with the property of possessing periodic orbits of the same period, around the origin, also in the original coordinates. See [5] for an analysis, via NF techniques, of systems of this type.

Corollary 2. *With the same notations as in Corollary 1, if* $\mathrm{Ker}(\mathcal{A}) \cap \mathrm{Ker}(\mathcal{B}) = \{0\}$, *then* f *can be linearized (possibly with a non-convergent NT), but, if* g *is analytic, there is a convergent NT which takes* f *into* $\mathrm{Ker}(\mathcal{A})$, *and another convergent NT which takes* f *into* $\mathrm{Ker}(\mathcal{B})$. *Also, if* g *is linear,* $g = Bx$, *then the linearizing transformation is convergent.*

It can be noted that, in fact, in all the above cases, Condition A is automatically satisfied. The last statement in Corollary 2 comes from the first part of Lemma 5, indeed $g = Bx$ is also a symmetry for the NF $\widehat{f}$, or, in other words, $\widehat{f}$ in $\mathrm{Ker}(\mathcal{A})$ implies $\widehat{f} \in \mathrm{Ker}(\mathcal{A}) \cap \mathrm{Ker}(\mathcal{B})$ and then the NF $\widehat{f}$ is necessarily linear. Let us also remark that the condition $\mathrm{Ker}(\mathcal{A}) \cap \mathrm{Ker}(\mathcal{B}) = \{0\}$ clearly follows from the requirement that $\mathcal{I}_A^* \cap \mathcal{I}_B^*$ is trivial, i.e. that $\mathcal{I}_A^* \cap \mathcal{I}_B^*$ contains only the constants.

We now give an example of a fourth-order DS $\dot{x} = f(x)$ in R^3 to which Corollary 2 can be applied.

`Example 1`. With $x \equiv (x, y, z) \in R^3$, consider the following DS:

$$\begin{aligned} \dot{x} &= x + a_1 x^3 y + b_1 x y^2 z \\ \dot{y} &= -\,3y + a_2 x^2 y^2 + b_2 y^3 z \\ \dot{z} &= 9z + a_3 y^2 z^2 + b_3 y^2 z^2 \end{aligned}$$

where a_i, b_i are arbitrary constants. A symmetry for this DS is actually a linear one, and is given by the VF

$$X_g = Bx \cdot \nabla = x\partial_x - 2y\partial_y + 4z\partial_z \ .$$

It can be easily verified that all the above assumptions are satisfied, in particular we have $\mathrm{Ker}(\mathcal{A}) \cap \mathrm{Ker}(\mathcal{B}) = \{0\}$, even if both $\mathrm{Ker}(\mathcal{A})$ and $\mathrm{Ker}(\mathcal{B})$ have infinite dimension, and even if neither the eigenvalues of A nor of B belong to a Poincaré domain. We can then conclude that this VF f can be linearized by a convergent transformation.

3. Convergence and Symmetries: A General Result

We now want to study the case that Condition A is *not* satisfied by the VF f, and to show that, under suitable conditions, the convergence of the NT of f can be equally ensured. These conditions are strictly related to the assumption that the given VF f, or the DS

$$\dot{x} = f(x) = Ax + F(x)$$

admits some *analytic* symmetry g (we refer obviously to *non-trivial* symmetries: i.e. g not proportional to f).

One of the first results in this direction was given by Markhashov [229] (his proof, however, appeared to be not complete [70]). Recently, Bruno and Walcher [73] obtained a remarkable result for the case of 2-dimensional problems (see Corollary 3).

Some recent results [87–89] on the problem of the convergence of NT, related to and generalizing the above mentioned ideas [73,229], are presented in this section and the next (see also [334] for a related approach focusing on the structure of the symmetry algebra). Here, the M_j are a basis in the space of the matrices commuting with A (clearly, one of these can be taken to be equal to A).

Theorem 4. *Given the analytic VF f, let us write its NF, according to Lemma 4, in the form*

$$\widehat{f} = Ax + \alpha(x)Ax + \sum{}' \mu_j(x) M_j x \equiv Ax + \alpha(x)Ax + \widehat{F}_1(x) \qquad (4)$$

where $\alpha(x), \mu_j(x) \in \mathcal{I}_A^*$ and (here and in the following) $\sum'$ is the sum extended to the matrices $M_j \neq A$. Assume $\widehat{F}_1(x) \neq 0$ (otherwise Condition A is sufficient to ensure convergence of the NT), and

i) *assume that f admits an analytic symmetry*

$$g = Bx + G(x) \qquad \text{such that} \qquad B = aA$$

where a is a (possibly vanishing) constant;

ii) *assume that the equation*

$$\{\widehat{F}_1, S\} = 0 \tag{5}$$

for the unknown power series

$$S = S(x) = \sum{}' \nu_j(x) M_j x \qquad \text{with} \qquad \nu_j(x) \in \mathcal{I}_A^* \qquad \text{and} \qquad (DS)(0) = 0 \tag{6}$$

has only the trivial solution

$$S = c\widehat{F}_1(x) \qquad\qquad (c = \text{constant}) \ .$$

Then f can be put into NF by means of a convergent NT.

Proof (a sketch, see [89] for details). First of all, if $a = 0$ in assumption i), one can consider, instead of g, the symmetry $g' = f + g$ having linear part Ax; it is then not restrictive to assume $a = 1$, i.e. $B = A$. Once f is put into NF $\widehat{f}$, the symmetry g will become a (possibly formal) symmetry $\widetilde{g}$

$$\widetilde{g} = Ax + \beta(x)Ax + \sum{}' \nu_j(x) M_j x \qquad \text{with} \qquad \beta(x), \nu_j(x) \in \mathcal{I}_A^* \ .$$

Evaluating, term by term, the bracket $\{\widehat{f}, \widetilde{g}\} = 0$ expressing the symmetry condition in the new coordinates, one is left with

$$\begin{aligned}\Big(\sum{}'_j \mu_j(x) M_j x \cdot \nabla\beta - \sum{}'_j \nu_j(x) M_j x \cdot \nabla\alpha\Big) Ax \, + \\ + \Big\{\sum{}'_j \mu_j(x) M_j x \ , \ \sum{}'_k \nu_k(x) M_k x\Big\} \ = \ 0 \ .\end{aligned} \tag{7}$$

Now, in this equation, the bracket $\{\ \cdot\ ,\ \cdot\ \}$ produces, through the matrix commutators $[M_j, M_k]$, only terms proportional to $M_j x$ (and not to Ax: this can be easily seen in a basis in which A is diagonal), therefore the terms appearing into the round and the curly brackets are separately zero. The last bracket has just the form $\{\widehat{F}_1, S\}$, and therefore assumption ii) gives $S = c\widehat{F}_1$, i.e. $\nu_j(x) = c\mu_j(x)$. From the vanishing of the term in round brackets in (7) one obtains similarly $\beta(x) = c\alpha(x)$, and then either $\widetilde{g} = c\widehat{f}$, which is impossible because $g \neq cf$, or

$$\widetilde{g}(x) \ = \ Ax \ .$$

This means that the transformation which puts $f(x)$ into $\widehat{f}(x)$ transforms $g(x) = Ax + G(x)$ into $\widetilde{g}(x) = Ax$, therefore the symmetry $g(x)$ satisfies Condition A and there is a convergent transformation which puts $g(x)$ into NF. Under this convergent transformation $f(x)$ is transformed into NF $\widehat{f}$, as a consequence of the last part of Lemma 5. ⊙

Remark 2. One can see that the assumption `ii)` of Theorem 4 implies that the NF $\widehat{f}$ admits only linear symmetries Lx. For a practical point of view (see the examples below), it is quite simpler to verify the property `ii)`.

`Example 2`. Consider a 3-dimensional analytic DS, with $x \equiv (x_1, x_2, z) \in \mathbf{R}^3$,

$$\dot{x} = f(x) = Ax + F(x) \qquad \text{with} \qquad A = \mathrm{diag}(1, 1, -2) \ ,$$

and assume that $f(x)$ possesses the linear SO_2 symmetry generated by $Lx \cdot \nabla$ where

$$L = \begin{pmatrix} 0 & 1 & 0 \\ -1 & 0 & 0 \\ 0 & 0 & 0 \end{pmatrix} \ ,$$

i.e. f is "equivariant" under rotations in the plane (x_1, x_2). Putting $r^2 = x_1^2 + x_2^2$, this implies that $F(x)$ must be written in the form

$$F(x) = \varphi_0(r^2, z)Ax + \varphi_1(r^2, z)Ix + \varphi_2(r^2, z)Lx \ ,$$

where I is the identity matrix in $\mathbf{R}^3$. If the functions $\varphi_0, \varphi_1, \varphi_2$ have the following form

$$\varphi_0 = 0, \qquad \varphi_1 = a_1 r^2 z + a_2 z^3 + \text{higher order terms}, \qquad \varphi_2 = b\varphi_1 \ ,$$

where a_1, a_2, b are constants $\neq 0$, then the DS admits also the nonlinear symmetry

$$G(x) = r^2 z(I + bL)x \ .$$

Notice that the assumption $a_2 \neq 0$ ensures that this DS is *not* a NF, and that the above symmetry is not trivial. Then assumption `i)` of Theorem 4 is satisfied. Now, the NF of the above DS must be of the form

$$\widehat{f} = Ax + \alpha(r^2 z)Ax + \mu_1(r^2 z)Ix + \mu_2(r^2 z)Lx \tag{8}$$

where α, μ_1, μ_2 depend only on $\kappa = r^2 z$, as a consequence of Lemma 5 (i.e., the equivariance under SO_2 is preserved), and of Lemma 4 (the resonance condition). We have to look for the solutions S of the equation $\{\widehat{F}_1, S\} = 0$, where the unknown S can be written

$$S = \nu_1(x)Ix + \nu_2(x)Lx \qquad \text{with} \qquad \nu_1(x), \nu_2(x) \in \mathcal{I}_A^* \tag{9}$$

[it is easy to see that no other matrices can appear in the r.h.s. of (8-9)], and where ν_1, ν_2 must be functions only of the two functionally independent quantities $x_1^2 z, x_1 x_2 z \in \mathcal{I}_A$. The condition $\{\widehat{F}_1, S\} = 0$ gives the first-order system of linear partial differential equations

$$\begin{aligned} (\mu_1 x + \mu_2 Lx) \cdot \nabla\nu_1 &= \nu_1 x \cdot \nabla\mu_1 \\ (\mu_1 x + \mu_2 Lx) \cdot \nabla\nu_2 &= \nu_1 x \cdot \nabla\mu_2 \ . \end{aligned}$$

Observing that $Lx \cdot \nabla\mu_1 = 0$ and $Ax \cdot \nabla\mu_1 = 0 = x \cdot \nabla\mu_1 - 3z\partial_z\mu_1$ and the same for ν_1, i.e. $x \cdot \nabla\nu_1 = 3z\partial_z\nu_1$, it is not difficult to show that the unique solution is just $S = c\widehat{F}_1$, then also the assumption ii) in Theorem 4 is satisfied, and therefore there is a convergent NT. It can be noted that in the calculations above it is essential that μ_1 and μ_2 are both $\neq 0$, and this is in fact guaranteed by the normalizing procedure: Indeed, at the lowest order, the resonant terms r^2zIx and r^2zLx are orthogonal to z^3Ix and z^3Lx (with respect to the standard scalar product [12,121,190] introduced in the vector space of homogeneous polynomials, where the homological operator $\mathcal{A}$ is defined), then at the first step of the normalization procedure the resonant terms are not changed, i.e. one has $\mu_1 = a_1r^2z + \ldots$ and $\mu_2 = a_1b\, r^2z + \ldots$; and then – at any subsequent step of the iteration – the lower order terms are not altered.

4. Convergence and Symmetries: A Special Case

In the particular case that $\widehat{F}_1(x)$ in Theorem 4 has the special form

$$\widehat{F}_1(x) = \mu(x)Mx$$

(with M not proportional to A), then assumption ii) of this theorem is actually equivalent to saying that: there are no common (analytic, formal or fractional) constants of motion of the two linear problems

$$\dot{x} = Ax \qquad \text{and} \qquad \dot{x} = Mx\ .$$

Indeed, if we assume that there is some $\kappa = \kappa(x) \in \mathcal{I}^*_A \cap \mathcal{I}^*_M$, then $S = \kappa(x)Ax \neq c\widehat{F}_1$ would satisfy $\{\widehat{F}_1, S\} = 0$. Note, incidentally, that one would also get in this case $\kappa(x) \in \mathcal{I}^*_{\widehat{f}}$. The converse is easily obtained by explicit calculations. This case has been considered in [87,88]. We can then state the two following theorems:

Theorem 5. *Assume that:*

i) *the DS admits an analytic symmetry $g = Bx + G(x)$ where the matrix B is proportional to A, $B = aA$ (including the case $a = 0$, but $G \neq aF$);*

ii) *once in NF, the DS takes the form*

$$\dot{x} = Ax + \widehat{F}(x) = Ax + \alpha(x)Ax + \mu(x)Mx \qquad\qquad \alpha(x),\ \mu(x) \in \mathcal{I}^*_A \tag{10}$$

where M is a matrix commuting with A but not proportional to A;

iii) *the two linear problems $\dot{x} = Ax$ and $\dot{x} = Mx$ do not admit analytic nor formal nor fractional common constants of motion.*

Then, the DS can be put in NF by means of a convergent normalizing transformation.

Theorem 6. *Instead of* i) *in Theorem* 5 *assume that the DS admits $\ell \geq 1$ analytic symmetries $g_j = B_jx + G_j(x)$, such that no linear combination of them is proportional to f, where the matrices B_j are linearly independent, and where ℓ is precisely the number of the linearly independent* linear *symmetries admitted* *by the DS once in NF. Then, with the conditions* ii),iii) *as in Theorem* 5, *the same conclusion holds.*

We can also state the following result, based on similar arguments, which is analogous to the Markhashov theorem [229].

Theorem 7. *Assume that the DS possesses a finite number $\ell \geq 1$ of analytic symmetries $g_j = B_j x + G_j(x)$, where all the matrices B_j are linearly independent. Assume also that, once in NF, the DS admits exactly ℓ linearly independent (possibly formal) symmetries. Then, there exists a converging normalizing transformation for f.*

The proof of these results are essentially based on the following steps [87,88]: 1) show that the DS admits a symmetry $g_0 = Ax + G(x)$ (with the same linear part as the DS), and recall that $g_A = Ax$ is a symmetry for the DS in NF; 2) show that there are no constants of motion for the NF, and – as a consequence – that $g_A = Ax$ is the unique symmetry admitted by this NF, having Ax as linear part; 3) apply, as in the proof of Theorem 4, the Condition A to the symmetry g_A, obtained under the transformation

$$g_0 = Ax + G(x) \to g_A = Ax$$

deducing that this transformation is convergent, and conclude that the same transformation also takes the DS into NF.

Theorem 7, however, appears to be less convenient for practical uses, in fact it requires the knowledge of *all* the symmetries of the NF. From this point of view, Theorems 5 and 6 (and Theorem 4) may be more useful and really applicable. In fact, the greatest apparent difficulty is that, in general, one does not know – a priori – the NF, therefore, it seems to be impossible to check if the assumptions of these theorems (or even Condition A) are verified by the NF. However, as already seen in Example 2 and also confirmed by the Examples 5,6, other symmetry properties of the VF may provide, once again, decisive help on this point.

Before discussing some applications and examples for Theorem 5, together with some related comments on the relevance of the role of symmetries in this context, let us give a few quite simple remarks that can be useful in the real applications.

Remark 3. If all the eigenvalues of A are non-degenerate, the number ℓ of the linearly independent linear symmetries for the NF satisfies $1 \leq \ell \leq (n-1)$. Indeed, if the eigenvalues of A are distinct, then there is basis where the matrices A, B and M_j are all diagonal, and the number of independent matrices commuting with A is precisely n. Among these, at least Ax is a symmetry; and at least Ix, where I is the identity matrix, is certainly not a symmetry (unless the DS is linear). Indeed, for any constant of motion μ of $\dot{x} = Ax$ one has $x \cdot \nabla\mu_j \neq 0$, whereas, with the notations of (3), $\{Ix, \widehat{f}\} = \sum_j (x \cdot \nabla\mu_j) M_j x$, and this would be zero only if $x \cdot \nabla\mu_j \neq 0$ for each j.

Remark 4. In the case of non-degenerate eigenvalues of the matrix A, the linear symmetries are clearly commuting. Note, however, that one may construct nonlinear symmetries that do not commute even if their linear parts are commuting (Example 3). Instead, in the case of degenerate eigenvalues, the symmetries of the DS do not commute, in general, as shown in Example 4 .

`Example 3`. Let $x \equiv (x_1, x_2, x_3, x_4) \in \mathbf{R}^4$, and consider the DS in NF

$$\dot{x} = (1 + x_1^2 x_2)Ax$$

where $A = \mathrm{diag}(1, -2, 3, -3)$. It is immediately verifiable that it admits the two following symmetries g_1 and g_2

$$g_1 = B_1 x \quad \text{where} \quad B_1 = \mathrm{diag}(1, -2, \sqrt{2}, -\sqrt{2})$$

$$g_2 = (1 + x_1^3 x_4)B_2 x \quad \text{where} \quad B_2 = \mathrm{diag}(1, -2, 0, -3) \ ,$$

which do not commute although $[B_1, B_2] = 0$.

`Example 4`. An interesting example of DS with degenerate eigenvalues and non-commuting symmetries is provided by the "quaternionic" $SU(2)$ case. Let again $x \in \mathbf{R}^4$, and consider the "realification" of the standard representation of $SU(2)$ [91,92,210], acting in $\mathbf{R}^4 \approx \mathbf{C}^2$, generated by real (4×4) matrices H_i, $i = 1, 2, 3$, such that $[H_i, H_j] = \varepsilon_{ijm} H_m$. There are then real (4×4) matrices K_j, $j = 0, \ldots, 3$, such that $K_0 = I$, $[K_i, K_j] = \varepsilon_{ijm} K_m$ for $i, j, m \neq 0$ (i.e. K_i, $i = 1, 2, 3$ generate a representation of $SU(2)$ as well), and commuting with the H_j, $[K_j, H_i] = 0$. Thus a DS of the form

$$\dot{x} = \sum_{i=1}^{3} \kappa_i(r^2) K_i x + \kappa_0(r^2) x \quad \text{where} \quad r^2 = (x, x)$$

admits the "quaternionic" symmetry described by the three non-commuting symmetries $H_i x$.

We now show that we can recover from Theorem 5 the following result by Bruno and Walcher [73] for 2-dimensional problems.

Corollary 3 [73]. *If a 2−dimensional DS, with semisimple linear part A having eigenvalues λ_1, λ_2 satisfying $k_1\lambda_1 + k_2\lambda_2 = 0$ (with k_1, k_2 nonnegative relatively prime integers not simultaneously zero), admits an analytic symmetry g (with no other assumptions), then it can be normalized by a convergent transformation.*

Indeed, if $B \neq aA$, then there exists a linear combination $c_1 g + c_2 f$ giving another symmetry of the DS such that its linear part $c_1 B + c_2 A = I$, then the DS can be linearized by a convergent NT, according to the Poincaré-Siegel theorem. If $B = aA$, observing that the NF has necessarily the form of (10) with $M = I$, then if $\mu = 0$ Condition A is satisfied. If instead $\mu \neq 0$ Theorem 5 can be applied, and we can conclude that in any case the NT is convergent.

Let us come back to Example 2, and assume now that, as in that example, the DS admits not only the nonlinear symmetry $g(x)$, but also a symmetry generated by some linear VF $Lx \cdot \nabla$, and that $g(x)$ satisfies

$$\{g, Lx\} = 0 \ . \tag{11}$$

In the NT the symmetry is conserved step by step [190], therefore $g(x)$ also will be transformed, once f is in NF $\widehat{f}$, into some $\widetilde{g}(x)$ which is symmetric under Lx: $\{\widetilde{g}, Lx\} = 0$. From this remark we see that, in order to verify assumption **iii)** of Theorem 5, it is sufficient to look for the common constants of motion of Ax and Mx *only in the set* of those scalar functions $\kappa = \kappa(x)$ which are left invariant by Lx. In other words, we can conclude that if the set of these simultaneously invariant functions is trivial, i.e. it contains only constant numbers

$$\mathcal{I}_A^* \cap \mathcal{I}_M^* \cap \mathcal{I}_L^* = \mathbf{R} \ , \tag{12}$$

then no other nonlinear symmetries are allowed, and $\widetilde{g}(x)$ becomes necessarily $\widetilde{g}(x) = Ax$. Then, using similar arguments as above, the convergence of the NT is guaranteed. In conclusion, we can state

Theorem 8. *Assume that f admits a symmetry g as in assumption* i) *in Theorem* 5, *and also a symmetry generated by a linear VF $Lx \cdot \nabla$ such that in addition* (11) *and* (12) *are satisfied. Then, if $\widehat{F}$ has the form* (10), *the NT is convergent. The result can be trivially extended to the case that f admits an algebra of (more than one) symmetries $L_k x \cdot \nabla$.*

Example 5. The same as Example 2, here with $b = 0$ and $a_1, a_2 \neq 0$. In the NF now $\mu_1 = a_1 r^2 z + \ldots \neq 0$, but μ_2 may be zero. If $\mu_2 \neq 0$, Theorem 4 can be applied. If $\mu_2 = 0$, then $\widehat{F}$ has the form (10); on the other hand, it is clear that no SO_2-invariant analytic functions are simultaneously constants of motion of $\dot{u} = Au$ and $\dot{u} = u$, then all assumptions of Theorem 8 are satisfied, and the NT is convergent.

Another (quite general) example, is the following.

Example 6. Let $n = 2m$ and, putting $x \equiv (u_1, \ldots, u_m, v_1, \ldots, v_m) \in \mathbf{R}^{2m}$; assume that a Lie group Γ acts "diagonally" on both the m-dimensional spaces of the vectors u and v through the same linear representation $\mathcal{D}$: i.e. $u \to u' = \mathcal{D}u$, $v \to v' = \mathcal{D}v$, where $\mathcal{D}$ is an absolutely irreducible representation (so the only matrix commuting with $\mathcal{D}$ is a multiple of the identity). Consider then a DS of the following form

$$\dot{x} = f(x) = Ax + F(x) \tag{13}$$

where

$$A = \begin{pmatrix} 0 & I_m \\ -I_m & 0 \end{pmatrix} \tag{13'}$$

and assume that $F(x)$ admits the symmetries $B_i x$, where B_i are the (matrix representatives in the direct sum $\mathcal{D} \oplus \mathcal{D}$ of the) Lie generators of this group Γ (the linear part Ax fulfills this symmetry requirement, so that the full DS (13) admits this symmetry). We shall choose, for instance, $m = 3$, $\Gamma = SO(3)$ and $\mathcal{D}$ its fundamental representation, and assume that the DS is

$$\dot{x} = Ax + p(x)x \tag{14}$$

where the scalar function p (cf. the example in [73]) is a homogeneous polynomial of degree $2k$ built with the quantities $u^2 = (u, u)$, $v^2 = (v, v)$ (where $(\cdot, \cdot)$ stands for the scalar product in $\mathbf{R}^3$). It is then easy to verify that the NF of the DS (14) must take the form

$$\dot{x} = Ax + \alpha Ax + \mu x \tag{15}$$

where α and μ are functions of $r^2 = u^2 + v^2$ only. On the other hand, one can see that the following vector function (with vanishing linear part)

$$g = (u^2 + v^2)^k x \tag{16}$$

generates a non-trivial nonlinear analytic symmetry for the DS (14), and that there are no common constants of motion for (15), as requested by iii) in Theorem 5. Therefore, the convergence of the normalizing transformation is ensured by this theorem. It is important to notice that, if our problem would not possess the symmetry Γ, the NF (15) would contain many other terms in its r.h.s., and that it is precisely the presence of the symmetry that forces the NF to contain only A and the identity, and allows us to apply the argument concerning the constants of motion. This seems to confirm the conjecture [73] that the presence of a "sufficient" number of symmetries may be an essential request in order to guarantee the convergence of a normalizing transformation.

5. Convergence in the Case of Hamiltonian Problems

The problem of the convergence of the NT in the case of Hamiltonian problems, even if similar to the analogous problem for DS, exhibits some special features. First of all, let us recall that the crucial Condition A for DS (Sect. 1 of this chapter) acquires in the Hamiltonian case a particular form, called by Bruno Condition H [66,67]. In the simplest case of eigenvalues lying on a straight line containing the origin in the complex plane, this condition becomes

Condition H. The NF $\widehat{H}$ of the Hamiltonian H is a series of the quadratic part H_0.

In the more general case, Condition H has a more complicated expression [66,67], but for our purposes we can state this condition in the above quite restrictive form, in analogy with Condition A for the DS. Note in particular that the first term of the expansion of $\widehat{H}$ is, in any case, precisely H_0.

Also, we explicitly assume, as for the DS, that the Condition ω is satisfied. Thus, one has

Theorem 9 (Bruno). *If a given Hamiltonian H satisfies Conditions H and ω, then H can be put into NF $\widehat{H}$ by means of a convergent NT.*

Unfortunately, Theorems 4-8 cannot be transformed into analogous results for the Hamiltonian case. Indeed a Hamiltonian DS never satisfies the crucial hypothesis, i.e. assumption `ii)` of Theorem 4. Let indeed $H = H(x)$, with $x \equiv (q,p) \in \mathbf{R}^{2m}$, be an analytic Hamiltonian and

$$\dot{x} = J\nabla H$$

be the associated DS. Writing $H = H_0 + H_R$, as in Chap. V, we have clearly $Ax = J\nabla H_0$ and $F(x) = J\nabla H_R$, and the requirement that $F(x)$ is in NF, $\widehat{F}(x) = J\nabla \widehat{H}_R$, now becomes the requirement that H_0 is a constant of motion of $\widehat{H}_R$ (Chap. V and [17,247]). Then, (5) in Theorem 4 always admits non-trivial solutions of the form

$$S = \eta(H_0)\widehat{F}_1$$

for any (regular) function η of H_0. Similarly, in the case of Theorem 5, there are always some obvious constants of motion of the problem in NF: precisely H_0 and $\widehat{H}$ itself (and functions thereof). Also, in any case, one has $\mathrm{Ker}(\zeta) \neq \{0\}$, indeed it contains, together with H_0 itself, all the constants of motion of H_0 (Chap. V). For the same reason, a similar difficulty arises in the presence of symmetries: if H admits a symmetry S, then in any case $\mathrm{Ker}(\zeta) \cap \mathrm{Ker}(\omega) \neq \{0\}$, in fact, $\mathrm{Ker}(\zeta) \cap \mathrm{Ker}(\omega)$ contains the simultaneous constants of motion of H_0 and S_0, and then, in any case, it contains – at least – the functions of H_0 and S_0. An additional difficulty appears for what concerns Theorem 6. In fact, it is true that H_0 is a symmetry of the NF $\widehat{H}$ (Theorem 1, Chap.V), just as Ax is a symmetry for the VF $\widehat{F}$ in NF, but one cannot be sure in this case that there is an analytic symmetry of H which is transformed into H_0 when H is transformed into NF.

So, to obtain some result on convergence of NT based on symmetry properties as in Sects. 3-4 of this chapter, one must resort to some stronger assumptions, along the lines of Theorem 3, e.g.:

Theorem 10. *Assume that the analytic Hamiltonian H admits an analytic symmetry $S = S_0 + \sum S_j$, with S_0 semisimple and satisfying Condition ω. Assume also that the eigenvalues of S_0 lie on a straight line containing the origin and that there is a coordinate transformation taking S into the quadratic part S_0, or – more in general – that S satisfies Condition H. Then there is a convergent transformation which takes H into an analytic Hamiltonian $\widetilde{H}$ which is in NF with respect to S_0 (i.e., $\widetilde{H} \in \mathrm{Ker}(\omega)$ and then $\widetilde{H}$ is a function of the constants of motion of S_0). In particular, if the above symmetry S is such that $S_0 = cH_0$ (with $c \neq 0$ and $S \neq cH$), then there is a convergent NT which puts H into an analytic NF $\widehat{H}$.*

Another result about convergence property of NT, based on the presence of symmetries of the Hamiltonian, is provided by the following theorem, obtained first by Rüssman [280] for the case of $\ell = 2$ degrees of freedom using Poincaré-Siegel conditions on convergence (Theorem 1) rather than Theorems 9-10 above, and then generalized by Vey [324] (see also [17]).

Theorem 11 (Russman-Vey). *If the analytic Hamiltonian H, with ℓ degrees of freedom, admits ℓ analytic constants of motion in involution $F^{(a)} = F_0^{(a)} + \sum_j F_j^{(a)}$ with the quadratic parts of the form*

$$F_0^{(a)} = \frac{1}{2} \sum_r \kappa_{ar}(q_r^2 + p_r^2) \qquad (a = 1, \dots, \ell) \tag{17}$$

and such that $\det(\kappa_{ar}) \neq 0$, *then the B transformation converges.*

Some other results in the same problem have been given by Ito [193–195], who considered the case of non-resonant eigenvalues (in the Hamiltonian sense) of the quadratic part H_0 of the Hamiltonian, and also the case of non-normal Hamiltonians (in the sense seen in Chap. V) admitting a single resonance. Calling H_S the "S part" of the Hamiltonian H, and ζ_s the associated homological operator, with the notations and definitions of Chap. V, his results can be summarized in the following Theorem [193–195]. Note in particular that (18,19) below agree with our Theorem 7 in Chap. V, taking $F^{(1)} = H$ and $\beta_1 = \zeta$.

Theorem 12 (Ito). *If the analytic Hamiltonian H, with ℓ degrees of freedom, admits ℓ functionally independent analytic symmetries $F^{(a)}$ in involution, and the eigenvalues of the quadratic part H_0 are all distinct and not resonant, then the B transformation is convergent. The same is true even if H_0 is not normal and even if there is just one resonance between two eigenvalues of the "S part" H_S of H. In this case, the B transformation takes H into NF with respect to H_S, i.e.*

$$H \to \widehat{H} \in \mathrm{Ker}(\zeta_s) \tag{18}$$

and also all symmetries $F^{(a)}$ of $\widehat{H}$ are in NF with respect to H_S, i.e. they satisfy

$$\widehat{F}^{(a)} \in \mathrm{Ker}(\zeta_s) \ . \tag{19}$$

Remark 5. The above results imply that the integrability near a non-resonant (or single-resonant) equilibrium point is equivalent to the existence of an analytic canonical B transformation.

The following example gives a simple case of a resonant Hamiltonian in $\ell = 2$ degrees of freedom, admitting an independent analytic symmetry. According to the above theorems, there is a convergent B transformation which puts this into NF.

Example 7. Consider the following Hamiltonian, not in NF,

$$H = \frac{1}{2}\left(x_1^2 + x_2^2 + p_1^2 + p_2^2\right) + \frac{1}{4}\left(x_1^2 + p_2^2\right)^2 = H_0 + H_R$$

It admits the quadratic symmetry

$$S = S_0 = x_1 x_2 + p_1 p_2$$

which, together with H_0, satisfies Theorem 12. Therefore the B transformation putting H into NF will be convergent, and, according to Theorem 1 and Theorem 2 of Chap. V, once in NF, this $\widehat{H}$ will be a function of H_0 and S_0 only.

VII. Invariant Manifolds

In this chapter we want to analyze the relations between the existence of symmetries $Y_s \in \mathcal{G}_f$ of a DS $\dot{x} = f(x)$ and the presence of manifolds invariant under the dynamical flow. Our interest will be mainly devoted to the properties of *center manifolds* (CM) [104,173,190,279,315]. It is well known that the reduction of the dynamics to the CM is a fundamental step in the analysis of the DS. Also, we assume that f may also depend on some real parameter λ, in order that bifurcation problems can be considered within this scheme (see Sect. 5 of this chapter for some examples). Our approach permits us to extend properties, well known in standard equivariant bifurcation theory [105,169,278,287,288,314], to the case of general LP symmetries. In particular, we can extend classical results obtained by means of Lyapunov-Schmidt projection to the case of bifurcation equations obtained by means of CM reduction [93,97]. The non-uniqueness (in general) of the CM also leads to some interesting features related to analyticity requirements and perturbation expansions. This aspect will be considered, in particular, in terms of the Poincaré-Dulac NF reduction and of the Shoshitaishvili theorem [104,297,298] on the topological equivalence of DS.

1. Some Preliminary Results on Flow Invariant Manifolds

Let us consider a DS

$$\dot{x} = f(x) = Ax + F(x) \tag{1}$$

admitting a symmetry $s = s(x)$. Putting, as usual,

$$X_f = f^i(x)\frac{\partial}{\partial x_i} \equiv f \cdot \nabla \quad ; \quad Y_s = s^i(x)\frac{\partial}{\partial x_i} \equiv s \cdot \nabla \ , \tag{2}$$

the well known symmetry condition

$$\{f, s\} = [X_f,\ Y_s] = 0 \tag{3}$$

is then satisfied.

We have the following basic Lemmata.

Lemma 1. *Any manifold W which is invariant under the dynamical flow is transformed by the symmetries Y_s into a flow-invariant manifold.*

Proof. Let $h(x) = 0$ be the equation describing the manifold W, then W is flow-invariant iff $X_f \cdot h\big|_{h=0} = 0$, where X_f is the generator of the dynamical flow. From (3), one has immediately

$$X_f(Y_s \cdot h)\big|_{h=0} = 0 \tag{4}$$

which implies the flow-invariance of the manifolds W_ε obtained by applying to W the transformation generated by Y_s (here, $\varepsilon \in R$ parametrizes the finite group action, and $W_0 \equiv W$, i.e., $h = 0$). ⊙

Let us now denote by E_0 and E_1 the subspaces spanned by the (generalized) eigenvectors of A having eigenvalues σ_i^0 with Re $\sigma_i^0 = 0$, so that E_0 will be the center eigenspace for the DS and, respectively, by E_1 the subspace of the (generalized) eigenvectors corresponding to all other eigenvalues of A. Performing a linear change of coordinates, and denoting by u the vectors in E_0 and by v those in E_1, the matrix A can be put in block-diagonal form and the problem (1) becomes

$$\begin{aligned} \dot u &= A_0 u + U(u,v) \\ \dot v &= A_1 v + V(u,v) \ . \end{aligned} \tag{5}$$

Writing a symmetry for this problem in the form

$$Y_s = s(x)\partial_x = \sum_{r=1}^{k} \varphi_r(u,v)\partial_{u_r} + \sum_{p=k+1}^{n} \psi_p(u,v)\partial_{v_p} \equiv \varphi\partial_u + \psi\partial_v \ , \tag{6}$$

we have the following results.

Lemma 2. *With the above notations, one has, for all indices r,s,p,q*

$$\frac{\partial\psi_p(0,0)}{\partial u_r} = \frac{\partial\varphi_s(0,0)}{\partial v_q} = 0 \qquad (r,s = 1,\dots,k \ ; \ p,q = k+1,\dots,n) \ . \tag{7}$$

Proof. Writing explicitly condition (3) with the above notations, one has in particular (sum over repeated indices)

$$\varphi_r \frac{\partial V_p}{\partial u_r} + \psi_q \frac{\partial V_p}{\partial v_q} = U_r \frac{\partial \psi_p}{\partial u_r} + V_q \frac{\partial \psi_p}{\partial v_q} \ .$$

Differentiating with respect to u_s, and putting $u = v = 0$, one finds, using also the property $s(0) = 0$,

$$(A_1)_{pq} \frac{\partial\psi_q(0,0)}{\partial u_s} = \frac{\partial\psi_p(0,0)}{\partial u_r}(A_0)_{rs} \ ,$$

or $A_1 D = D A_0$ where D is the constant matrix $D_{pr} = \partial\psi_p(0,0)/\partial u_r$, which implies $D = 0$. An identical argument shows $\partial\varphi_s(0,0)/\partial v_q = 0$. ⊙

Lemma 3. *Let $v = w(u)$ be the equation describing any $(n-k)$–dimensional smooth manifold W tangent at the origin to the subspace $v = 0$:*

$$w(0) = 0 \qquad \partial_u w(0) = 0 \ .$$

Then all transformations $Y_s = \varphi\partial_u + \psi\partial_v$ which satisfy the conditions (7) and $s(0) = 0$ change this manifold into a manifold which is still tangent at the origin to $v = 0$.

Proof. This immediately follows by observing that the manifold W, which is given by the equation $h \equiv v - w(u) = 0$, is changed under Y_s according to $Y_s \cdot h = \psi - \partial_u w \cdot \varphi$, but this transformation, thanks to (7), preserves the tangency at the origin. ⊙

It is interesting that (in contrast with Lemma 1) the result of Lemma 3 is essentially "geometrical" (not "dynamical"), in fact we do not need here the hypothesis that W is an invariant manifold under a dynamical flow, nor that Y_s is a symmetry.

An obvious consequence obtained combining all the above conclusions (in short, symmetries preserve flow-invariance of manifolds and their tangency property) is the following.

Theorem 1. *All symmetries $Y_s \in \mathcal{G}_f$ leave invariant (i.e. map into themselves, not necessarily pointwise) both the stable and unstable manifolds W of the DS, i.e.*

$$Y_s : W \to \mathrm{T}W \ . \tag{8}$$

Also, the symmetries Y_s transform any CM into a CM (possibly the same one).

The symmetry invariance of stable and unstable manifolds can actually be inferred directly from Lemma 1. In fact, they are flow-invariant and uniquely defined, being characterized by the time-exponential behavior of the trajectories in them, so that the symmetries, being time-independent, cannot map them into different invariant manifolds. The situation is not so simple (and in fact a weaker result is true) in the case of the CM, which is in general not unique, and where the dynamics is more complicated and interesting. The non-uniqueness of the CM, and the related possibility that a symmetry map a CM into a different CM, lead to some non-trivial features within the present context, which will be briefly discussed in Sects. 4,5 of this chapter.

It should be remarked that the last statement of the above theorem means, in the language of fluid dynamics [51,228], that a symmetry transformation cannot connect *slow* and *fast* modes.

To conclude this section, it can be interesting to check, from the point of view of the symmetry transformations, at least in one especially simple situation, one of the peculiar properties of CM. It is known in fact that, if W is a CM, there is an open subset Ω, with $\Omega \cap W \neq \emptyset$, such that if a solution $f(t) \in \Omega$, $\forall t \in R$, then $f(t)$ belongs to a CM [104,190,279,315]. Assume here for simplicity that the CM is unique, the closure $\bar{\Omega}$ is compact and that a solution $f(t)$ actually belongs $\forall\ t \in R$, to $W \cap \Omega$. Applying the transformation generated by a symmetry $Y_s \in \mathcal{G}_f$, one obtains a $f_\varepsilon(t)$ which is again a solution. But, $\forall \delta > 0$, $\exists \varepsilon$ such that $\sup_{t \in R} |f_\varepsilon(t) - f(t)| < \delta$, due to the smoothness of Y_s. Then if δ is small enough, $f_\varepsilon(t) \in \Omega$ and then also $f_\varepsilon(t) \in W$ for the mentioned property of the CM, which confirms the previous result that W is mapped into itself by the symmetries $Y_s \in \mathcal{G}_f$.

For some related considerations, and the possible extension of these properties to the infinite-dimensional case, see [75,155,316].

2. Reduction to a Center Manifold

We now examine more closely the dynamics. With the above notations,

$$x \equiv (u,v) \in R^n \quad ; \quad u \in E_0 \quad v \in E_1 \tag{9}$$

where in particular u span the k−dimensional center eigenspace E_0, let us choose a CM W, described by the $(n-k)$ equations

$$v = w(u) \tag{10}$$

and put

$$z = v - w(u) \qquad z \in R^{n-k} \ . \tag{11}$$

Then the original DS becomes from (5)

$$\begin{aligned} \dot{u} &= A_0 u + \widetilde{U}(u,z) \\ \dot{z} &= A_1 z + Z(u,z) \ , \end{aligned} \tag{12}$$

where $\widetilde{U}(u,z) = U(u,v(u,z))$ etc., and $Z(u,0) = 0$, as a consequence of the flow-invariance of the CM $z=0$. Note that if the CM (10) is C^k, then the DS (12) will be only C^k although the original DS was C^∞ (or even analytic).

The symmetry Y_s (6) then becomes

$$\widetilde{Y}_s = \widetilde{\varphi}(u,z)\partial_u + (\widetilde{\psi} - \partial_u w \cdot \widetilde{\varphi})\partial_z \equiv \Phi(u,z)\partial_u + \Psi(u,z)\partial_z \ . \tag{13}$$

Then we can say that $\Psi(u,0)$ is the term "pushing away" from the CM. It satisfies the equation, obtained by writing the symmetry condition (3) and restricting it to $z=0$:

$$A_1\Psi(u,0) + \partial_z Z(u,0) = \partial_u \Psi(u,0)\big(A_0 u + \widetilde{U}(u,0)\big) \ . \tag{14}$$

We also have

Lemma 4. *The CM* (10) *is left invariant by the symmetry* (6) *if and only if*

$$\psi(u,w(u)) = \big(\partial_u w(u)\big)\cdot\varphi\big(u,w(u)\big) \qquad i.e. \qquad \Psi(u,0) = 0 \ . \tag{15}$$

An immediate but important consequence of the above arguments concerns the symmetries inherited by the dynamics restricted to the CM. In fact, we have the following result.

Theorem 2. *If the symmetry $Y_s \in \mathcal{G}_f$ admitted by the DS leaves invariant a given CM $v = w(u)$, then the DS restricted to this CM, i.e.*

$$\dot{u} = A_0 u + \widetilde{U}(u,0) = A_0 u + U\big(u,w(u)\big) \tag{16}$$

possesses the symmetry generated by

$$\widehat{Y}_s = \Phi(u,0)\partial_u = \varphi(u,w(u))\partial_u \ . \tag{17}$$

If, in particular, the CM is unique, then this is true for any symmetry admitted by the DS.

This is a useful result. Indeed, if the CM is unique, this proposition ensures that the DS once reduced to the CM, automatically inherits the symmetries of the original problem. Then this result can be viewed as the counterpart of the Sattinger theorem in bifurcation theory [287,288]. This theorem, in fact, stated in the context of equivariant bifurcation theory, relates the symmetries of the original problem to those of the reduced problem obtained by means of Lyapunov-Schmidt projection.

Some further discussion is needed for the case when the CM is not unique. In fact, this non-uniqueness is related to the presence of non-analytic terms, and thus in some sense this non-uniqueness can be "controlled" by some analyticity arguments.

3. Normal Forms

In the previous section we have introduced the new variable z (11) in such a way that the CM can be expressed in the form $z = 0$. An alternative way of obtaining a DS in a form where "automatically" one CM is given by a "flat" manifold is provided by the Poincaré-Dulac procedure of reducing the DS to normal form. We can give a direct proof (cf. also [121]) of the following result. We consider here for simplicity the case when A is semisimple (the case of non-semisimple A is completely analogous).

Theorem 3. *If the DS is written in the Poincaré-Dulac normal form*

$$\dot{x} = Ax + F(x) \tag{18}$$

where all nonlinear terms $F(x)$ are resonant with the linear part A, then, with the notations of (5), *one CM is given by*

$$v = 0 \ .$$

Proof. As shown in Chap. IV, the resonant terms have the form $F = K(\kappa(x))x$ with $AK = KA$, and $\kappa = \kappa(x)$ are the constants of motion of the problem $\dot{x} = Ax$. Writing as in (9) $x \equiv (u, v)$ and $A = \begin{pmatrix} A_0 & 0 \\ 0 & A_1 \end{pmatrix}$, this implies $K = \begin{pmatrix} K_0 & 0 \\ 0 & K_1 \end{pmatrix}$ and thus the DS has the form

$$\begin{aligned} \dot{u} &= A_0 u + K_0\big(\kappa(u, v)\big)u \\ \dot{v} &= A_1 v + K_1\big(\kappa(u, v)\big)v \ . \end{aligned} \tag{19}$$

The conclusion then easily follows after observing that (19) shows that $v = 0$ is a flow-invariant manifold. ⊙

Note that it is easy to obtain all the constants of motion $\kappa(u,v)$ of the *linear* system $\dot{x} = Ax$, and then conclude also the following.

Theorem 3′. *If in particular in the DS* (19) *the real part of* all *the eigenvalues* σ_i *of* A_1 *is negative (or positive), then the only smooth constants of motions* $\kappa(u,v)$ *may actually* depend only on the variables u. *The same is true if there is no linear dependence over* **N** *between the eigenvalues of* A_1, *i.e. no relation* $m_1\sigma_1 + \ldots + m_n\sigma_n = 0$, *where* $m_i \geq 0$ *are non-negative integer numbers (not all zero). In these cases, the time evolution of the variables* u *turns out to be independent of the variables* v:

$$\dot{u} = A_0 u + K_0\big(\kappa(u)\big)u \ . \tag{20}$$

This result generalizes Theorem 3 in [121] (cf. also eq. $(5)_1$ of [121], and [189,309]).

4. Shoshitaishvili Theorem and Center Manifolds

The non-uniqueness of the CM is related in an essential way to non-perturbative effects [16,104,119,190,217]. The investigation of non-analytic CM by means of symmetry methods would require the introduction of non-analytic symmetries. It is clear indeed that analytic symmetry transformations map analytic manifolds into manifolds with the same property. Let us comment on this point. Assume, for instance, that the DS has the following simple form

$$\begin{aligned} \dot{u} &= A_0 u + U(u) \\ \dot{v} &= A_1 v \ , \end{aligned} \tag{21}$$

where the eigenvalues of A_0 have vanishing real parts, and here A_1 is a constant matrix whose eigenvalues have nonzero real parts. Note that actually the Shoshitaishvili theorem [104,297,298] ensures the local *topological* equivalence of any DS to a system of this simple form.

A CM for this DS is clearly $v = 0$. If we assume that another CM is given by $v = w(u)$, then by putting, as in Sect. 2, $z = v - w(u)$ we obtain

Lemma 5. *The time-evolution of the variable* z *is the same as that of the variable* v:

$$\dot{z} = A_1 z \ .$$

Proof. We have

$$\dot{z} = A_1 v - \partial_u w \cdot \dot{u} \ ,$$

but the invariance of the CM $h \equiv v - w(u) = 0$ under the dynamical flow X_f, i.e. $X_f \cdot h\Big|_{h=0} = 0$, implies that the function $w(u)$ satisfies

$$A_1 w = \partial_u w \cdot \dot{u} = \partial_u w \cdot \big(A_0 u + U(u)\big) \ ;$$

therefore

$$\dot{z} = A_1 v - A_1 w = A_1 z \ .$$

$\odot$

We now want to evaluate explicitly, in the particularly simple form (21), how a symmetry can transform a CM into a different CM. Recalling from Sect. 3 of this chapter that the component $\Psi(u,0)$ of the symmetry is the term "pushing away" from the CM, we have from (14) that, in the special case of (21), the term $\Psi(u,0)$ satisfies the equation

$$A_1 \Psi(u,0) = \partial_u \Psi(u,0 \cdot \left(A_0 u + U(u,0)\right) \ . \tag{22}$$

Consider for instance the DS, with $x \equiv (u_1, u_2, v) \in R^3$, and putting $r^2 = u_1^2 + u_2^2$

$$\begin{aligned} \dot{u}_1 &= -\, u_1 r^2 \\ \dot{u}_2 &= -\, u_2 r^2 \\ \dot{v} &= -\, v \ . \end{aligned} \tag{23}$$

This DS admits a family of CM, given by

$$v = \mathcal{F}\Big(\frac{u_2}{u_1}\Big) \mathrm{e}^{1/2r^2} \ , \tag{24}$$

where $\mathcal{F}$ is an arbitrary regular function of the constants of motion u_2/u_1. A symmetry admitted by this DS is the linear one (it is just the rotation) generated by

$$Y = u_2 \frac{\partial}{\partial u_1} - u_1 \frac{\partial}{\partial u_2} \ , \tag{25}$$

which clearly leaves invariant the flat (and analytic) CM

$$v = 0 \tag{26}$$

and also, e.g., the other CM

$$v = \mathrm{e}^{-1/2r^2} \ . \tag{27}$$

But the DS admits the other (non-analytic) symmetry

$$Y' = r^2 (u_1 \frac{\partial}{\partial u_1} + u_2 \frac{\partial}{\partial u_2}) - \mathrm{e}^{-1/2r^2} \frac{\partial}{\partial v} \tag{28}$$

that *does* leave invariant this CM (27) but *not* the flat one (26). Instead, the other CM in the family (24)

$$v = \frac{u_1^2}{r^2} \mathrm{e}^{-1/2r^2} \tag{29}$$

is not left invariant by the two above symmetries, but this other symmetry can be introduced

$$Y'' = u_1^2 (u_1 \frac{\partial}{\partial u_1} + u_2 \frac{\partial}{\partial u_2}) - \mathrm{e}^{-1/2r^2} \frac{\partial}{\partial v} \tag{30}$$

leaving it invariant.

Exercise 1. Verify that the manifolds given in (24) are in fact CM for the DS (23), and that the VF

$$Y = \mathcal{F}_0\left(\frac{u_1}{u_2}\right) r^2 (u_1 \frac{\partial}{\partial u_1} + u_2 \frac{\partial}{\partial u_2}) + \mathcal{F}_1\left(\frac{u_1}{u_2}\right) v \frac{\partial}{\partial v}$$

where $\mathcal{F}_0$, $\mathcal{F}_1$ are arbitrary regular functions, are symmetries for this DS. For any given CM in the family (24), choose the functions $\mathcal{F}_0$, $\mathcal{F}_1$ in such a way that the CM is left invariant by this symmetry.

The analysis of non-analytic manifolds and non-analytic symmetries would lead to problems that are beyond the goal and the scope of the present work and we can refer the reader to [16,97,104,119,190,217]. Anyway, as the above example and the exercise indicate, we can say that – at least under suitable mild assumptions – for any CM there is a symmetry which leaves it invariant, and conversely any symmetry leaves invariant a CM. This allows us to suitably extend Theorem 2 also to the case of non-unique CM.

5. Some Examples

Consider first the following DS, with $x \equiv (u_1, u_2, v) \in R^3$, according to the notations of Sect. 3 of this chapter,

$$\begin{aligned} \dot{u}_1 =& u_1^2 f_1(\kappa_1, \kappa_2) \\ \dot{u}_2 =& u_1(u_1^2 - v) f_2(\kappa_1, \kappa_2) + f_3(\kappa_1, \kappa_2) + u_1^2(x_1^2 - v) f_1(\kappa_1, \kappa_2) \\ \dot{v} =& (v - u_1^2) f_2(\kappa_1, \kappa_2) + 2u_1^3 f_1(\kappa_1, \kappa_2) \ , \end{aligned} \tag{31}$$

where f_i are arbitrary smooth functions of the quantities

$$\kappa_1 = u_1^2 - v \ ; \qquad \kappa_2 = u_2 - u_1^3 + u_1 v \ , \tag{32}$$

with the conditions

$$\lim_{x \to 0} f_2(\kappa_1, \kappa_2) = \ell \neq 0; \qquad \lim_{x \to 0} f_3(\kappa_1, \kappa_2) = \lim_{x \to 0} \partial_x f_3 = 0 \ . \tag{33}$$

It is easy to verify that a symmetry admitted by (31) is

$$Y = u_1^2 \partial_{u_1} + u_1^2(u_1^2 - v)\partial_{u_2} + 2u_1^3 \partial_v \tag{34}$$

and the (unique) CM for this DS is

$$v = u_1^2 \ . \tag{35}$$

Note that, as this and the following examples clearly indicate, the invariance of the CM under a symmetry may also be useful for explicitly finding the CM itself. Indeed, the expression defining the CM must be a function only of the quantities $\kappa_\alpha(x)$ which are left invariant by the symmetry Y, i.e.

$$Y \cdot \kappa_\alpha = 0 \ . \tag{36}$$

In this example, these quantities are precisely given by (32). According to the discussion in Sect. 3 of this chapter, the restriction of the DS (31) to the CM is

$$\begin{aligned} \dot{u}_1 &= u_1^2 f_1(0, u_1) \\ \dot{u}_2 &= f_3(0, u_2) \end{aligned} \tag{37}$$

which inherits from (31,34) the symmetry

$$\widehat{Y} = u_1^2 \partial_{u_1} \tag{38}$$

admitted in fact by (37).

Let us introduce another class of DS, again with $x \equiv (u_1, u_2, v) \in R^3$ (the extension to $n > 3$ dimensions is straightforward, see [93,97]). Let $\kappa = \kappa(u_2, v)$ be an arbitrary smooth function, satisfying only the condition $\frac{\partial \kappa}{\partial v} \neq 0$, and put $\chi = \chi(u_2, v) = \partial_{u_2}\kappa / \partial_v \kappa$. Consider the DS

$$\begin{aligned} \dot{u}_1 &= u_1 \ f - u_2 \ g \\ \dot{u}_2 &= u_2 \ f + u_1 \ g \\ \dot{v} &= - u_2 \ f \ \chi - u_1 \ g \ \chi + \frac{h}{\partial_v \kappa} \end{aligned} \tag{39}$$

where f, g, h are arbitrary smooth functions of the quantities

$$r^2 = u_1^2 + u_2^2 \qquad \text{and} \qquad \kappa = \kappa(u_2, v)$$

(and possibly also of one or more real parameters λ, to include the case of bifurcation problems). Then, using (3), it is easy to verify that the above DS (33) admits the symmetry generated by

$$Y = u_2 \partial_{u_1} - u_1 \partial_{u_2} + u_1 \chi \partial_v \ . \tag{40}$$

If κ and h in (39) satisfy

$$\lim_{u_2, v \to 0} \partial_{u_2} \kappa = 0 \qquad \text{and} \qquad \lim_{u_2, v \to 0} \frac{h(r^2, \kappa)}{v \ \partial_v \kappa} = \ell < 0 \tag{41}$$

then the origin $x_0 = 0$ is an equilibrium point for the problem (39), with an attractive 2-dimensional center manifold W tangent at $x_0 = 0$ to the center eigenspace E_0 ($v = 0$). If the functions $f(\lambda, r^2, \kappa)$, $g(\lambda, r^2, \kappa)$ (with $\lambda \in R$) satisfy also

$$f(0,0,0) = 0, \quad \partial_\lambda f(0,0,0) \neq 0, \quad g(0,0,0) \neq 0 \tag{41'}$$

then the standard conditions for the appearance of Hopf bifurcation are met.

As remarked above, the invariance under the symmetry (40) implies that the CM must be a function of the quantities $\kappa_\alpha(x)$ defined in (36), which are in this case just r^2 and κ.

E.g., let us choose in (39)

$$\kappa = v\,\mathrm{e}^{-u_2} \quad \text{and} \quad f = h = -\kappa + r^2 \tag{42}$$

leaving $g = g(r^2, \kappa)$ arbitrary. This gives $\chi = -v$, $h/\partial_v \kappa = -v + \mathrm{e}^{u_2} r^2$, and the nonlinear symmetry

$$Y = u_2 \partial_{u_1} - u_1 \partial_{u_2} - u_1 v \partial_v \ . \tag{43}$$

It is easy to explicitly obtain the CM, which is given by the surface

$$v = w(u_1, u_2) = \mathrm{e}^{u_2} r^2 \tag{44}$$

and to verify also that the symmetry inherited, according to Sect. 3 of this chapter, by the problem reduced to the CM is the rotation in the (u_1, u_2) plane, generated by

$$\widehat{Y} = u_2 \partial_{u_1} - u_1 \partial_{u_2} \ . \tag{45}$$

It can be interesting to perform explicitly the calculations indicated in Sect. 3 to transform this problem to its CM. By introducing the variable

$$z = v - \mathrm{e}^{u_2} r^2$$

(just as in (11)), we obtain with the above choice (42) the problem

$$\begin{aligned} \dot{u}_1 &= - u_1 z \mathrm{e}^{-u_2} - u_2 g \\ \dot{u}_2 &= - u_2 z \mathrm{e}^{-u_2} + u_1 g \\ \dot{z} &= z(-1 + u_1 g + 2r^2 - u_2 z \mathrm{e}^{-u_2}) \end{aligned} \tag{46}$$

where $g = g(r^2, r^2 + z\mathrm{e}^{-u_2})$. We then see from (46) that $z = 0$ is in fact an invariant attractive solution (the CM, as expected!), and also that *only on the* CM $z = 0$ does the system exhibit the rotation symmetry (45).

Let us point put that, in this example, the symmetry inherited by the reduced problem is a *linear* symmetry (45), even if the original symmetry (43) is *not* a linear one. This may be a remarkable occurrence: in fact, if the original problem has *no* linear symmetries, then standard (linear) equivariant theory cannot predict any symmetry constraints on the form of the reduced equations. This may be important e.g. when dealing with Hopf bifurcation theory, where symmetry properties play a relevant role, as shown by the following example.

Let us choose in (39) $\kappa = v - u_2^2$, $h = -\kappa$, $f = \lambda - r^2$, with arbitrary $g(\lambda, r^2, \kappa)$ (satisfying in particular all conditions (41,41′)). The resulting bifurcation problem

$$\begin{aligned} \dot{u}_1 &= u_1\ (\lambda - r^2) - u_2\ g(\lambda, r^2, \kappa) \\ \dot{u}_2 &= u_2\ (\lambda - r^2) + u_1\ g(\lambda, r^2, \kappa) \\ \dot{v} &= - v + u_2^2(1 + 2(\lambda - r^2)) + 2u_1 u_2\ g(\lambda, r^2, \kappa) \end{aligned} \tag{47}$$

admits the symmetry

$$Y = u_2 \partial_{u_1} - u_1 \partial_{u_2} - 2u_1 u_2 \partial_v \ . \tag{48}$$

It is easy to verify that at the critical point $\lambda_0 = 0$, the CM is given by

$$v = u_2^2 \tag{49}$$

(this manifold is a flow-invariant manifold also for $\lambda \neq 0$), and the reduced bifurcation equations exhibit the (linear) rotation symmetry (45). Now, a SO_2-symmetric Hopf periodic solution to (47) bifurcates at $\lambda > \lambda_0 = 0$, along the lines

$$r^2 = u_1^2 + u_2^2 = \lambda, \qquad v = u_2^2$$

and with frequency $\omega(\lambda) = g(\lambda, \lambda, 0)$.

VIII. Further Normalization

In this volume we have considered the standard normalization procedure of Poincaré, that depends on generating functions (for the changes of coordinates) h_m which are obtained as solutions of the relative homological equations. But these solutions are not unique whenever the homological operator, which will be denoted here as $\mathcal{L}_0$, has a nontrivial kernel. Functions h_m differing for an element $\delta h_m \in \text{Ker}(\mathcal{L}_0)$ produce the same effect on F_m, but not on higher order terms F_k $(k > m)$. Correspondingly, the normal form is not unique, and the unfolding of NFs – which can be described as the most general *resonant* system [i.e. a system with nonlinear part in $\text{Ker}(\mathcal{L}_0^+)$] – is exhaustive, but in general not minimal. That is, the Poincaré-Dulac theory provides a list of NFs such that any system can be (formally) reduced to one of the NFs of the list, but we are not guaranteed that the NFs are pairwise not conjugated.

Having a minimal – or at least less redundant – classification would be of interest for different reasons. First, by having a reduced classification we have less cases to study when we want to completely describe the behavior of systems with a given linear part (however, as mentioned above, this should be taken with some care due to the problem of convergence of the normalizing transformation). Second, a reduced classification will also correspond to simpler normal forms: When we have to resort to a perturbation argument or to numerical integration in order to study the behavior of normal forms, disposing of some nonlinear terms would allow us to obtain better precision (either by considering higher order perturbations, or by allowing for shorter time-steps in numerical integration at equal machine time). This advantage should be quite clear e.g. in the setting considered in the planar example considered below (where results already known to Siegel and Moser for Hamiltonian systems are extended to the non-Hamiltonian case).

Thus it would be useful to provide a reduced classification of NFs, i.e. a list of NFs (say, corresponding to a given linear part A) in which the redundancies in the Poincaré-Dulac classification, or at least some of them, have been eliminated.

This problem has been considered by several authors [18,19,61,62,65,206, 209,210,317], mainly in the language of Lie algebras filtration. Indeed, several theoretical results exist (see e.g. the section on "further normalization" in [317]), and it is actually also possible to define a *unique NF* [18]. However, all these results – to the best of our knowledge – are difficult to implement.

In this chapter we discuss a procedure of "higher order normalization" (which for the formal case $m = \infty$ we call "renormalization") which is completely algorithmic, and not more difficult to implement than the standard Poincaré-

Dulac normalization (we follow [148,149]). We will also discuss how the presence of symmetries modifies this procedure.

This procedure is nothing else than a direct generalization – or, more modestly, iteration – of the standard one. The main "new" ingredient will be to make use of the freedom in the choice of the generating functions h_k for the near-identity coordinate transformations that comes from the nontriviality of the kernel of $\mathcal{L}_0$, and a control of higher-order effects in this transformation.

Not surprisingly if one considers that, as mentioned above, previous results were obtained in the framework of Lie algebras, it is convenient to consider Lie-Poincaré [47,240] rather than standard Poincaré transformations (details are given below). However, we will *not* discuss the Lie-theoretic side of the procedure described here since the Lie-theoretic frame has been discussed with great clarity by other authors and because we wanto to focus attention on how this higher order normalization is completely in the original frame of the Poincaré approach.

1. Higher-Order Terms in Poincaré Transformations

As mentioned above, the "renormalization" procedure will be based on using the freedom in the choice of h_m in order to control – as much as possible – the effects of transformation on higher order terms. We will thus consider these here.

We deal, as usual, with a system of the form

$$\dot{x} \;=\; f(x) \;=\; \sum_{k=0}^{\infty} f_k(x) \tag{1}$$

where f_k is homogeneous of degree $(k+1)$, so that in particular $f_0(x)$ represents the linear part of (1), $f_0(x) = Ax$ with $A = (Df)(0)$, and consider near-identity changes of coordinates of the form

$$x = y + h_k(y) \qquad , \qquad h_k \in V_k \; . \tag{2}$$

We denote by Γ the Jacobian of the change of coordinates, i.e. ${\Gamma^i}_j = \partial h_k^i / \partial y^j$. Under the change of coordinates (2), our system (1) is transformed into

$$\dot{y} = [I + \Gamma]^{-1} \; f\left(y + h_k(y)\right) \; . \tag{3}$$

For y – and therefore x – small enough, $\Lambda = (I + \Gamma)^{-1}$ does surely exist, and we can write it in a power series as $\Lambda \equiv (I + \Gamma)^{-1} = \sum_{m=0}^{\infty} [(-1)^m (\Gamma)^m]$.

Similarly, we can expand $f_m(y + h_k(y))$ as a power series. We write $J = (j_1, ..., j_n)$, $|J| = \sum_i j_i$. With this multiindex notation, $\partial_J := \partial_1^{j_1} ... \partial_n^{j_n}$, and similarly $h_k^J := (h_k^1)^{j_1} ... (h_k^n)^{j_n}$. We define the operators $\Phi_h^r = (1/r!) \sum_{|J|=r} \left(h^J \cdot \partial_J \right)$ (representing all the partial derivatives of order $|J|$), and in terms of these the system (3) can then be written as

$$\dot{y} \;=\; \sum_{m=0}^{\infty} \sum_{r=0}^{\infty} \sum_{s=0}^{\infty} \left[(-1)^s \; \Gamma^s \Phi_{h_k}^{r-s} \right] \; f_m(y) \; . \tag{4}$$

Thus we see that in a Poincaré transformation with generator $h_k \in V_k$ each term f_m is transformed into a new $\widetilde{f}_m$ given by

$$\widetilde{f}_m = f_m + \sum_{p=1}^{[m/k]} \left[\sum_{s=0}^{p} (-1)^s \Gamma^s \Phi_{h_k}^{p-s} \right] f_{m-kp} , \tag{5}$$

where the square brackets in $[m/k]$ denote the integer part. One could, in principle, also obtain explicit formulas for terms of all degrees, but these become quickly too involved to be of practical use. We notice, however, that the terms of degree smaller than k are not changed at all, $\widetilde{f}_m = f_m \; \forall m < k$, and the terms of degree $k \leq m < 2k$ are changed according to

$$\widetilde{f}_{k+\nu} = f_{k+\nu} + [\Phi_h - \Gamma] \, f_\nu \quad (0 < \nu < k) . \tag{6}$$

A slightly different way of approaching Normal Forms is based on Lie – rather than Poincaré – transformations [47,240], as seen in previous chapters. In this case, the change of coordinates is given by the time-one action of a vector field H_k given by

$$H_k = h_k^i(x) \frac{\partial}{\partial x^i} \tag{7}$$

so that the change of coordinates is written as

$$\widetilde{x} := y = \left[e^{-\lambda H_k} x \right]_{\lambda=1} . \tag{8}$$

In this way, X is transformed into

$$\widetilde{X} = \left[e^{\lambda H_k} \, X \, e^{-\lambda H_k} \right]_{\lambda=1} , \tag{9}$$

and this can be explicitly computed by the Baker-Campbell-Haussdorf formula [215,240]. We will just recall the final result, i.e. that $\widetilde{X}$ can be written as $\widetilde{X} = \widetilde{f}^i(x)\partial_i$, with

$$\widetilde{f} = f + \{h, f\} + \frac{1}{2}\{h, \{h, f\}\} + \frac{1}{6}\{h, \{h, \{h, f\}\}\} + \ldots . \tag{10}$$

From (10) it is easy to derive formulas for the decomposition of $\widetilde{f}$ into homogeneous factors, i.e. for $\widetilde{f} = \sum_m \widetilde{f}_m$. We introduce the notation $\mathcal{H}(.) = \{h, .\}$, and with this we have

$$\widetilde{f}_m = \sum_{s=0}^{[m/k]} \frac{1}{s!} \mathcal{H}^s \left(f_{m-sk} \right) . \tag{11}$$

Note that we have written again $[m/k]$ for the integer part of (m/k), and defined $\mathcal{H}^0(f) = f$.

2. The Homological Operators

We will define a series of homological operators $\mathcal{L}_k$ associated with f. The usual homological operator, which we will denote in this section by $\mathcal{L}_0$ (and which was also denoted by $\mathcal{A}$ in previous chapters), will be the first of these. This definition will suit our way of proceeding, based on Poincaré-Lie transformations and thus on (10).

For $f \in V$, $f = \sum f_k$, we define the Lie operator $\mathcal{F} : V \to V$ associated with f as $\mathcal{F} = \{f, .\}$. Clearly we can write

$$\mathcal{F} = \sum_{k=0}^{\infty} \{f_k, .\} \equiv \sum_{k=0}^{\infty} \mathcal{L}_k . \tag{12}$$

The operators $\mathcal{L}_k = \{f_k, .\}$ defined in (12) are called the sequence of homological operators associated with f. The operator $\mathcal{L}_0$ coincides with the usual homological operator considered in Poincaré normal form theory. Notice that $\mathcal{L}_k : V_m \to V_{m+k}$. We also denote by $\mathcal{L}_{k,m}$ the restriction of $\mathcal{L}_k$ to V_m.

Note that when we operate a Poincaré transformation, the f_k – and thus the $\mathcal{L}_k$ – change. However, at each normalization stage we stabilize a new term f_s, and thus the corresponding $\mathcal{L}_s$.

It should be emphasized that linear combinations of the homological operators do not permit to describe (10), or (11), in full generality: they are only related to the first nontrivial term in (11). However, it will turn out that, in the procedure we employ in the following, a suitable choice of the h allows us to analyze iterated Poincaré-Lie transformations in terms of the $\mathcal{L}_k$ alone.

3. Non-uniqueness of Poincaré Normal Forms

In the Poincaré procedure, or the Lie-Poincaré we prefer to consider in this chapter, one has no need to keep track of the effect of the transformation generated by $h_k \in V_k$ on terms of higher order. Indeed, this will generate additional terms in V_s, in principle at all the higher orders $s > k$, but these can then be disposed of by the successive Poincaré transformations with generator h_s.

This point should be considered with some extra care. While the terms generated by $\mathcal{L}_0(h_s)$ are in $\mathrm{Ran}(\mathcal{L}_0) \cap V_s = \mathrm{Ran}\,(\mathcal{L}_{0,s})$, those appearing as "higher order terms" due to the transformation generated by h_k (with $k < s$) cannot be guaranteed to be (and in general, are not) in the same space. Thus, not all of these can then be eliminated by suitably choosing h_s.

However, this feature should not necessarily be seen as a drawback in the Poincaré procedure. If on the one hand this shows that we could introduce resonant terms that were not initially present in our system, on the other hand we could use the same mechanism to eliminate (some of the) resonant terms initially present, or generated as higher order terms by previous changes of coordinates. More generally, we could use these higher order effects of the Poincaré transformations to further normalize – we will use the term *renormalize* – the Poincaré normal form.

As mentioned in the introduction to this chapter, it is well known that the Poincaré normal form is by no means unique, i.e. that two different Poincaré normal forms can be conjugated. The idea of further normalization is also not new, and has been considered by several authors [18,19,209,210,317], mainly in the context of Lie algebras filtration. We mention, in particular, the work of Broer [61,62] and Broer and Takens [65], the "further normalization" of van der Meer [317], and the "unique normal forms" studied by Baider [18]. However, such an approach seems difficult to implement in the study of concrete dynamical systems. We want here to study the same problem from a direct point of view, i.e. by considering the properties of Poincaré transformations and the explicit higher order effects, see (5) and (11). This allows us to give a well defined algorithm, which – once a basis is chosen in each of the V_k – only requires linear algebraic computations. Moreover, these reduce to consideration of the action of (higher order) homological operators and to solution of relevant (higher order) homological equations, thus representing a natural and straightforward generalization of the Poincaré scheme.

4. Poincaré Renormalization

As a preliminary step, we have to introduce some abstract notation. We will consider some chains of spaces and operators, together with the chain of spaces $V_k \subset V$ and the chain of operators $\mathcal{L}_k : V \to V$ introduced above. These definitions will be based on a formal power series $f = \sum f_k$ (as was already the case for the $\mathcal{L}_k$).

We define the spaces $H^{(p)} \subseteq V$ by $H^{(0)} = V$ and, for $p \geq 1$,

$$H^{(p)} = \mathrm{Ker}(\mathcal{L}_0) \cap \ldots \cap \mathrm{Ker}(\mathcal{L}_{p-1}) = \bigcap_{s=0}^{p-1} \mathrm{Ker}\,(\mathcal{L}_s) \ . \tag{13}$$

It is obvious that $H^{(p+1)} \subseteq H^{(p)}$ (so that the $H^{(p)}$ realize a filtration of V).

We define then the operators $\mathcal{M}_k$ as the restriction of $\mathcal{L}_k$ to $\mathrm{Ker}(\mathcal{L}_0) \cap \ldots \cap \mathrm{Ker}(\mathcal{L}_{k-1})$, i.e. $\mathcal{M}_p = \mathcal{L}_p|_{H^{(p)}}$. Clearly,

$$\mathrm{Ker}\,(\mathcal{M}_p) = \bigcap_{s=0}^{p} \mathrm{Ker}\,(\mathcal{L}_s) = H^{(p+1)} \ . \tag{14}$$

Next, we define the spaces $F^{(p)}$ by $F^{(0)} = V$ and

$$F^{(p)} = \bigcap_{k=0}^{p-1} [\mathrm{Ran}\,(\mathcal{M}_k)]^{\perp} = \bigcap_{k=0}^{p-1} \mathrm{Ker}\,(\mathcal{M}_k^{+}) \ ; \tag{15}$$

hence the $F^{(p)}$'s satisfy $F^{(p+1)} \subseteq F^{(p)}$ (and realize a filtration of V, as was already the case for the $H^{(p)}$). Note that $F^{(p+1)} = F^{(p)} \backslash \left[\mathrm{Ran}\,(\mathcal{M}_p) \cap F^{(p)}\right]$.

We can also define projection operators for each of these spaces. We denote them by

$$\begin{aligned} \pi_s : V \to \mathrm{Ker}(\mathcal{L}_s) \quad &, \quad \widehat{\pi}_s : V \to \mathrm{Ran}(\mathcal{L}_s^+) \\ \chi_s : V \to \mathrm{Ker}(\mathcal{L}_s^+) \quad &, \quad \widehat{\chi}_s : V \to \mathrm{Ran}(\mathcal{L}_s) \; . \end{aligned} \tag{16}$$

It will be useful to define the composition of projection operators given by

$$\mu_s = \pi_{s-1} \circ \pi_{s-2} \circ \ldots \circ \pi_0 \qquad (s \geq 1) \tag{17}$$

(and $\mu_{s,k}$ as the restriction of this to V_k). Later on we will use the notation μ_0, that should be taken as the identity operator. We also consider the projection on the range of $\mathcal{M}_s$, i.e.

$$P_s : V \to \mathrm{Ran}\,(\mathcal{M}_s) \; . \tag{18}$$

In this way, we can redefine our spaces and operators by using the projection operators. In particular we have $H^{(s)} = \mu_s\,[V]$ and $\mathcal{M}_s = \mathcal{L}_s \circ \mu_s$.

Remark 1. Notice that we could consider the decomposition of the spaces and operators introduced above according to $V = \sum^{\oplus} V_k$, i.e. by considering their intersection with, and restriction to, the spaces V_k. As in each of the V_k (considered as vector spaces) we have a finite dimensional basis, the relevant operators – in particular, the $\mathcal{M}_{p,k} = \mathcal{M}_p\big|_{V_k}$ – can be written in matrix form by using these bases.

Thus, to implement the above remark, we define the intersections $H_k^{(p)} := H^{(p)} \cap V_k$. These satisfy

$$H_k^{(p+1)} = H_k^{(p)} \cap \mathrm{Ker}(\mathcal{L}_{p,k}) \; . \tag{19}$$

We can then also consider the finite dimensional operators $\mathcal{M}_{p,k} : H_k^{(p)} \to V_{k+p}$ given by the restriction of $\mathcal{L}_{p,k}$ (recall that this is the restriction of $\mathcal{L}_p$ to V_k) to $H_k^{(p)}$; clearly, $H_k^{(p+1)} = \mathrm{Ker}(\mathcal{M}_{p,k})$. The adjoints of these, defined of course with respect to the scalar product defined in the V_k and thus in V, satisfy $\mathcal{M}_{p,k}^+ : V_{k+p} \to H_k^{(p)}$. Finally, the intersections $F_k^{(p)} := F^{(p)} \cap V_k$ do also satisfy, for $k \geq p$,

$$F_k^{(p+1)} = F_k^{(p)} \cap \mathrm{Ker}(\mathcal{M}_{p,k-p}^+) \; . \tag{20}$$

We can use these considerations and the general formulas obtained for the higher order action of a Poincaré transformation to devise a scheme of Poincaré normalization which includes iterated normalization (renormalization) and leads to a simplified normal form unfolding. We choose to work in the Lie-Poincaré scheme. The reason for this preference will be clear in the following.

We sketch below the construction of such renormalized forms, first for orders two and three, and then give the general algorithm in terms of the abstract setting introduced in the beginning of this section.

It should be stressed that we give directly a convenient "open" algorithm, i.e. one which does not require to fix in advance the order to which we want to go to in the renormalization procedure. A slightly different approach, in which we fix this in advance, is maybe a more immediate extension of the Poincaré procedure and could therefore provide a better conceptual understanding (but is computationally not convenient). This is discussed below in the next section.

We start from a system like (1), and denote by $A = (Df)(0)$ the matrix corresponding to the linear part of f (which will not be modified by our transformations). We rewrite the system, for further reference, as

$$\dot{x} = \sum_{k=0}^{\infty} f_k^{(0)}(x) , \tag{21}$$

where the upper index on f_k (and similar ones from now on) refers to the number of Poincaré transformations of order k applied so far.

We can then, by means of the usual Poincaré transformation with generator $h_1^{(0)}$ chosen as a solution to the homological equation for $k = 1$, take the term $f_1^{(0)}$ into NF, i.e. transform it to

$$f_1^{(1)} = f_1^{(0)} - \mathcal{L}_0 \left(h_1^{(0)} \right) \in \operatorname{Ker} \left(\mathcal{L}_0^+ \right) \cap V_1 . \tag{22}$$

Let us now go on to consider the term in V_2 (which is now already changed, but we still denote it by $f_2^{(0)}$ for ease of notation). By the usual Poincaré procedure, i.e. by choosing $h_2^{(0)}$ as solution to the homological equation at order $k = 2$, this is changed into

$$f_2^{(1)} = f_2^{(0)} - \mathcal{L}_0 \left(h_2^{(0)} \right) \in \operatorname{Ker} \left(\mathcal{L}_0^+ \right) \cap V_2 . \tag{23'}$$

However, we know that we can still change this by a transformation with generator $h_1^{(1)} \in V_1$. It is clear that now this $h_1^{(1)}$ cannot be completely arbitrary, if we want to keep the already normalized term $f_1^{(1)}$ in its present – and satisfactory – form. Thus, to leave this untouched, we are obliged to require

$$h_1^{(1)} \in \operatorname{Ker}(\mathcal{L}_0) \cap V_1 . \tag{23''}$$

It is clear that in this way we can only eliminate terms which are in $\operatorname{Ran}(\mathcal{M}_1)$, where we have defined

$$\mathcal{M}_1 = [\mathcal{L}_1]_{\operatorname{Ker}(\mathcal{L}_0)} . \tag{24}$$

Indeed, notice that (23′) implies that in (11) we are reduced to $\widetilde{f}_2 = f_2 + \{h_1, f_1\}$. It is easy to see that this same remark, suitably generalized, will also hold in general for our proposed procedure. The transformations will be given by action of the series of homological operators (this represents the main explicit advantage, in the present discussion, of considering Lie transformations).

Thus, this second normalization of the term in V_2 will transform $f_2^{(1)}$ into

$$f_2^{(2)} = f_2^{(1)} - \mathcal{L}_1 \left(h_1^{(1)} \right) = f_2^{(1)} - \mathcal{M}_1 \left(h_1^{(1)} \right) , \tag{25}$$

and by suitable choice of $h_2^{(1)}$ we can have

$$f_2^{(2)} \in \left[\operatorname{Ker} \left(\mathcal{L}_0^+ \right) \cap V_2 \right] \cap \operatorname{Ker} \left(\mathcal{M}_1^+ \right) . \tag{26'}$$

If we denote by P_1 the operator of projection on $\mathrm{Ran}(\mathcal{M}_1)$, this "suitable choice" amounts to [here the pseudoinverse $\mathcal{M}_1^{-1}$ is only defined on $\mathrm{Ran}(\mathcal{M}_1)$]

$$h_1^{(1)} = \mathcal{M}_1^{-1} P_1 f_2^{(1)} . \tag{26''}$$

We could then repeat the same procedure for the term $f_3^{(0)}$, by applying Poincaré transformations with generators $h_3^{(0)}$ (solution of the homological equation at order $k = 3$), $h_2^{(1)}$, and $h_1^{(2)}$. This time, however, we should not only make sure that the $h_2^{(1)}$ does not modify the already established $f_2^{(2)}$, i.e. that

$$h_2^{(1)} \in \mathrm{Ker}(\mathcal{L}_0) , \tag{27'}$$

but also that $h_1^{(2)}$ does not modify neither the $f_2^{(2)}$ nor the $f_1^{(1)}$, i.e. that

$$h_1^{(2)} \in \mathrm{Ker}(\mathcal{L}_0) \cap \mathrm{Ker}(\mathcal{L}_1) . \tag{27''}$$

Again, these requirements will guarantee that in (11) only the term $\mathcal{L}_1(h_2^{(1)})$ and $\mathcal{L}_2(h_1^{(2)})$ are effectively giving a contribution to the transformed f_3.

The general scheme ofthe construction should be quite clear at this point, and we can now describe it in abstract terms with the notation introduced above.

Definition 1. *The dynamical system* $\dot{x} = f(x) = \sum f_k(x)$ *(the vector field* $X = f^i(x)\partial_i = \sum f_k^i(x)\partial_i$*; the formal power series* $f(x) = \sum f_k(x)$*) is in* Poincaré renormalized form *up to order* n *if* $f_k \in F_k^{(k)}$ *for all* $k \le n$.

Proposition 1. *Any dynamical system (vector field, formal power series) can be brought into Poincaré renormalized form up to any desired order* n *by means of a formal series of Lie-Poincaré transformations.*

Proof. We will prove constructively the above proposition by giving a well defined algorithm for the transformation to Poincaré renormalized form.

We operate sequentially for $k = 1, 2, ..., n$ in the following way. If $f_k^{(0)}$ is the term of order k after performing the required transformation at orders up to $k-1$, we then operate a series of Lie-Poincaré transformations with generators $h_k^{(0)}, h_{k-1}^{(1)}, ..., h_1^{(k-1)}$, where $h_p^{(s)} \in H_p^{(s)}$. This condition guarantees, see above, that at each step the transformation generated by $h_{k-p}^{(p)}$ will transform $f_k^{(p)}$ into

$$f_k^{(p+1)} = f_k^{(p)} - \mathcal{M}_p\left(h_{k-p}^{(p)}\right) . \tag{28}$$

The $h_k^{(0)}$ is chosen as the solution to the standard homological equation, i.e. as

$$h_k^{(0)} = \mathcal{L}_0^+ P_0 f_k^{(0)} \equiv \mathcal{M}_0^{-1} P_0 f_k^{(0)} . \tag{29}$$

The $h_{k-p}^{(p)}$ should be chosen as the projection on $H_{k-p}^{(p)}$ of the solution to the (higher order) homological equations

$$P_p f_k^{(p)} - \mathcal{M}_p\left(h_{k-p}^{(p)}\right) = 0 , \tag{30}$$

which means, explicitly,

$$h_{k-p}^{(p)} = \mu_p \circ \mathcal{M}_p^{-1} \circ P_p \left(f_k^{(p)} \right) \tag{31}$$

(with $\mu_0 = I$, the (29) is included in this formula as well).

Clearly, in this way we arrive in the end (i.e. after applying the procedure to terms of order $k = 1, 2, 3, ...n$) at a system

$$\dot{x} = f^*(x) = \sum_{k=0}^{\infty} f_k^*(x) \tag{32}$$

in which, for $k \leq n$,

$$f_k^*(x) = f_k^{(k)} \in F_k^{(k)} . \tag{33}$$

The proof is thus complete. ⊙

Remark 2. We would like to point out that one can also work following a different order: once we have fixed the order n up to which we want to put the system in normal form, we can first proceed with the usual Poincaré normalization, i.e. consider the transformations generated on the $h_k^{(0)}$, and corresponding to the action of $\mathcal{L}_0$, for $k = 1, ..., n$. We can then consider the transformations, generated by $h_k^{(1)}$, corresponding to the action of $\mathcal{L}_1$ (or more precisely of $\mathcal{M}_1$), i.e. "second-normalize the normal form", and so on. This will be illustrated in the next section.

Remark 3. We would also like to remark explicitly that, although we have preferred to avoid cumbersome notation, one could consider restriction of operators to the relevant V_k subspaces, and thus transform – by means of bases in each of the V_k – the above equations into algebraic ones. In order to do this, we should use the notation introduced after Remark 1, and perform our computations using explicitly the basis $e_{\mu,i}$ introduced in Sect. 1 of this chapter. In computer-assisted computations, it would be appropriate to proceed in this way.

5. Renormalization by Iterated Normalizations

We have discussed so far a convenient construction of the renormalized forms. In this section we discuss a slightly different construction. This is computationally less convenient, but it has the advantage that it literally amounts to repeating the Poincaré procedure several times, each time using a higher, stabilized at previous stages, homological operator $\mathcal{L}_k$ instead of the standard $\mathcal{L}_0$.

It should be pointed out that although the result we can prove in this way is the same, and although the unfolding of the renormalized form corresponding to the given linear part is exactly the same, by following the two algorithms we can obtain *different* renormalized forms.

We will use the same notation introduced above (as the reader will immediately notice, this discussion will be very similar to the previous one). Suppose now to have already transformed the initial system into its Poincaré-Dulac normal form, which we rewrite as

$$\dot{x} = f_0^{(0)}(x) + \sum_{k=1}^{\infty} f_k^{(1)}(x) \tag{34}$$

where $f_0^{(0)} = f_0$ and $f_k^{(1)} \in \mathrm{Ker}(\mathcal{L}_0^+) \cap V_k := F_k^{(1)}$.

We can now start to consider again the Lie-Poincaré transformations with generators $h_k^{(1)} \in V_k$ (the upper index reminds us that we have already performed one Poincaré normalization). In order to be sure that these do not take the term f_k (of the same order) out of normal form, we must require that $h_k^{(1)} \in \mathrm{Ker}(\mathcal{L}_0)$. In this way no term in V_k is produced, but we still have terms produced at higher orders, i.e. we are changing f_m for $m > k$.

Let us at first concentrate in particular on the effects at order $k+1$. The term f_{k+1} changes, due to (11), according to

$$f_{k+1} \rightarrow f_{k+1} - \mathcal{L}_1(h_k^{(1)}) . \tag{35}$$

This means that we can change $f_2^{(1)}$ into a

$$f_2^{(2)} = f_2^{(1)} - \mathcal{L}_1(h_1^{(1)}) \tag{36}$$

without changing $f_1^{(1)}$ (because of $h_1^{(1)} \in \mathrm{Ker}(\mathcal{L}_0)$, as we required above).

Note that in this way we do not generate non-resonant terms at higher orders, as the transformation operates by iterated brackets between functions in $\mathrm{Ker}(\mathcal{L}_0)$ [equivalently, commutators of vector fields which all commute with $X_0 = f_0(x)\partial_x$], see (10,11), and this set is obviously closed under the bracket operation. The same kind of considerations apply when considering higher order normalizations. This is strictly related to Lie algebra filtrations (mentioned in the introduction to this chapter), and is a substantial advantage of the Lie-Poincaré approach in this context.

Proceeding in this way for $k = 1, 2, \ldots$, we can, by a suitable choice of the $h_k^{(1)}$, change the $f_m^{(1)}$ with $m \geq 2$ and eliminate any term in the image of $\mathrm{Ker}(\mathcal{L}_0)$ under $\mathcal{L}_1$, i.e. in the range of $\mathcal{M}_1$. In this way we arrive to a "second normalized form" for our system:

$$\dot{x} = f_0^{(0)} + f_1^{(1)} + \sum_{k=2}^{\infty} f_k^{(2)} . \tag{37}$$

Notice that here $f_0^{(0)} \in V_0$ is still the original one, $f_1^{(1)} \in \mathrm{Ker}(\mathcal{L}_0^+)$ is the one obtained with the first normalization, and all the $f_k^{(2)}$ are in $\mathrm{Ker}(\mathcal{M}_1^+)$.

We can now repeat the same procedure by choosing $h_k^{(2)} \in \mathrm{Ker}(\mathcal{L}_0) \cap \mathrm{Ker}(\mathcal{L}_1)$ and concentrating on the effect of the Poincaré transformation generated by this on f_{k+2}, changing all terms with $k \geq 3$ and producing, in the end, a "third normalized form"

$$\dot{x} = f_0^{(0)} + f_1^{(1)} + f_2^{(2)} + \sum_{k=3}^{\infty} f_k^{(3)} , \tag{38}$$

and so on.

The general procedure is now clear. It is also clear that in this way we can, by operating n successive "higher order" normalizations, take the system into renormalized form up to order n. We will state the general result that can be obtained in this way as follows.

Definition 2. *The dynamical system $\dot{x} = \sum_{k=0}^{\infty} f_k$ (the vector field $X = f^i(x)\partial_i$, the formal power series $f(x) = \sum f_k(x)$) is said to be in* n-th normal form *if $f_k \in F_k^{(p)}$ for all k, with $p = \min(k,n)$. When this condition is satisfied for all n (with no upper limit), we say that the system is in* Poincaré renormalized form.

Notice that a system in n-th normal form is also in renormalized form up to order n.

Proposition 2. *Given a dynamical system $\dot{x} = \sum_{k=0}^{\infty} f_k^{(0)}$ (a vector field $X = f^i(x)\partial_i$, a formal power series $f(x) = \sum f_k(x)$), it is always possible to reduce it, by a sequence of n formal Poincaré normalizations, into the n-th normal form*

$$\dot{x} = \sum_{k=0}^{n} f_k^{(k)} + \sum_{s=n+1}^{\infty} f_s^{(n)} ,$$

where $f_k^{(k)} \in F_k^{(k)}$ for all $k \leq n$.

We can also say formally, considering infinite sequences of Poincaré normalizations, that a system can always be taken into renormalized form.

For a more thorough discussion of this renormalization procedure, we refer the reader to [149], which we have followed in this chapter. We will now present some examples.

6. Examples: Planar Vector Fields

We will consider, as a first and meaningful example [148,149], the unfolding of normal forms for vector fields in R^2 having a linear part $f_0(x) = Ax$ with

$$A = \begin{pmatrix} 0 & -1 \\ 1 & 0 \end{pmatrix} . \tag{39}$$

As is well known, the Poincaré normal forms corresponding to this can be written in the form

$$f(x) = Ax + \sum_{k=1}^{\infty} \left(x_1^2 + x_2^2\right)^k [a_k I + b_k A]\, x , \tag{40}$$

where the a_k, b_k are arbitrary real constants. Writing $r^2 = (x_1^2 + x_2^2)$, this reads

$$f_k(x) = \begin{cases} 0 & \text{for } k \text{ odd} \\ r^{2m}\left[a_m I + b_m A\right] x & \text{for } k = 2m. \end{cases} \tag{41}$$

We will see that the renormalized normal form unfolding is remarkably simpler. In order to do this, and to avoid trivial steps, we will consider the renormalized form of a system which is already in Poincaré normal form.

Thus, let us consider an f which has already been taken into Poincaré normal form, and proceed to its renormalization. We will consider only the non-trivial transformations (but keep the indices' notation introduced above).

First consider the term $f_2^{(0)}$. Since it is in normal form (and since for the same reason $f_1 = 0$, i.e. $\mathcal{L}_1 \equiv 0$), we cannot modify it by our algorithm, i.e. it will remain in the form given above, $f_2^{(2)} = f_2^{(0)} = r^2(a_1 I + b_1 A)x$.

We then have $f_3 = 0$, and $f_4^{(0)} = r^4(a_2 I + b_2 A)x$. The generator $h_2^{(2)}$ of the transformation $f_4^{(2)} = f_4^{(0)} - \mathcal{M}_2\left(h_2^{(2)}\right)$ must be, for $h_2^{(2)} \in \mathrm{Ker}(\mathcal{L}_0)$, of the form $h_2^{(2)} = r^2(\alpha I + \beta A)x$, and thus $\mathcal{M}_2\left(h_2^{(2)}\right) = 4\left(a_1\beta - b_1\alpha\right) Ax$. It is then clear that, unless $a_1 = b_1 = 0$, we can always choose α, β so that

$$f_4^{(2)}(x) = f_4^{(4)}(x) = r^4 a_2 x \ . \tag{42}$$

Now consider f_6. Now $h_4^{(2)}(x) = r^4(\alpha I + \beta A)x$ [again for $h \in \mathrm{Ker}(\mathcal{L}_0)$], and

$$\mathcal{L}_2\left(h_4^{(2)}\right) = \left[(-2a_1\alpha)I + (2a_1\beta - 4b_1\alpha)A\right] x \ . \tag{43}$$

Thus, if $a_1 \neq 0$, we can eliminate f_6 completely in this way.

It is quite easy to become convinced that, under the same condition $a_1 \neq 0$, the same holds for all the f_{2m}. Indeed, for $h_{2(k-1)}^{(3)}(x) = r^{2(k-1)}\left[\alpha I + \beta A\right] x$ [again, $h_{2(k-1)}^{(3)} \in \mathrm{Ker}(\mathcal{L}_0)$ requires this form for h], we have

$$\mathcal{L}_2\left(h_{2(k-1)}^{(3)}\right) = -\left[(2(k-2)a_1\alpha)\, I + (2(k-1)a_1\beta - 2b_1\alpha)\, A\right] x \ . \tag{44}$$

Thus, we can completely eliminate all the f_{2m}.

We have thus shown that:

Lemma 1. *If in the Poincaré normal form (40) for $f(x)$ the constant a_1 is non-zero, then the corresponding Poincaré renormalized form is given by*

$$f^*(x) = Ax + r^2\left(a_1 I + b_1 A\right) x + r^4 a_2 x \ ,$$

and thus its unfolding depends on three real parameters.

It is also possible to consider the degenerate situation where $a_1 = 0$. Actually one can obtain a rather general result, which we give here without proof [149]:

Lemma 2. *Let the vector formal power series $f : R^2 \to R^2$ be given by $f(x) = Ax + \sum_{k=1}^{\infty} r^{2k}\left[a_k I + b_k A\right] x$; let μ be the lowest number such that $a_\mu \neq 0$, and ν the lowest number such that $b_\nu \neq 0$, so that $f(x)$ can be written as*

$$f(x) = Ax + \sum_{k=\mu}^{\infty} r^{2k} a_k x + \sum_{k=\nu}^{\infty} r^{2k} b_k Ax . \tag{45}$$

Then, the Poincaré renormalized form of f up to any given order n is given by

$$f^*(x) = Ax + r^{2\mu} a_\mu x + r^{2\nu} \beta Ax + r^{4\mu} \alpha x , \tag{46}$$

where $a_\mu \neq 0$ is the same as in (45), and the α, β could (possibly, but not necessarily) vanish. In particular, if $\nu > \mu$, then $\beta = 0$.

Note that in the symplectic case, i.e. when all the a_k in (40) vanish, this corresponds to a classical result [299], recently extended to higher dimension [138].

It should be pointed out that the coefficients a_μ, α, β appearing in (45) have a natural "physical" meaning for the behavior and stability of solutions to the system under consideration. If we go to polar coordinates (r, ϑ), then $\dot{x} = f^*(x)$ reads

$$\begin{cases} \mathrm{d}r / \mathrm{d}t = r^{2\mu+1} (a_\mu + r^{2\mu} \alpha) \\ \mathrm{d}\vartheta / \mathrm{d}t = 1 + r^{2\nu} \beta \end{cases} . \tag{47}$$

Thus, if a_μ and α are both positive, the only limited solution is the trivial one, $x(t) \equiv 0$, which is unstable; for a_μ and α both negative, $x(t) \equiv 0$ is stable, and no proper periodic solution, nor solutions which are limited for $t \to -\infty$, exist.

For $(a_\mu \alpha) < 0$, there is a periodic solution, given by motion on the circle of radius r_0 given by

$$r_0 = |a_\mu / \alpha|^{1/(2\mu)} . \tag{48}$$

The motion on this circle is counterclockwise for $\beta > -1/r_0^{2\nu}$, i.e.

$$\beta > \left(\frac{\alpha}{a_\mu}\right)^{\nu/\mu} \equiv \beta_0 < 0 , \tag{49}$$

clockwise for $\beta < \beta_0$, and degenerate into a circle of fixed points for β exactly equal to β_0. Obviously, this solution is given by $\dot{\vartheta} = \omega_0$ with

$$\omega_0 = 1 + \left|\frac{a_\mu}{\alpha}\right|^{\nu/\mu} \beta . \tag{50}$$

As for the stability of this solution (when non-degenerate), it is easily seen from the first of (47) that we have a stable circle r_0 for $a_\mu > 0$ and $\alpha < 0$, and an unstable one in the case $a_\mu < 0$, $\alpha > 0$.

This result can be generalized to the case, again in R^2 with coordinates (x_1, x_2), with linear part $f_0(x) = Ax$ given by

$$A = \begin{pmatrix} 1 & 0 \\ 0 & -n \end{pmatrix} \tag{51}$$

with $n > 1$ an integer. This is discussed in some detail in [149] together with a number of simple three-dimensional examples.

The example considered above dealt with a system whose linear part corresponds to a semisimple matrix A. We do now want to consider the case in which this is not diagonalizable (i.e. A contains a nilpotent part), and we will actually consider the simplest case in which A is itself nilpotent:

$$A = \begin{pmatrix} 0 & 1 \\ 0 & 0 \end{pmatrix} . \tag{52}$$

In this case, obviously, $\mathcal{L}_0$ will not coincide with $\mathcal{L}_0^+$, and the work needed for implementing our procedure will be more complicated than in the cases previously analyzed.

It will turn out that in this case the reduction obtained with respect to the standard Poincaré normal form is rather weak.

Since we have to consider $\mathcal{L}_k^+$, we have to choose a scalar product. We will choose the Bargmann scalar product $(.,.)$ introduced in Chap. II. This choice will be understood in all the statements of this section.

We will again refer to [149] for proofs, and just give the results, referring to a standard (Poincaré) normal form expansion of the kind $f(x) = Ax + \sum_{k=2}^{\infty} f_k(x)$. We have first that:

Proposition 1. *If in the standard normal form expansion the quadratic term f_1 does not vanish, the system can always be written (formally) as*

$$\begin{pmatrix} y \\ 0 \end{pmatrix} + a_1 \begin{pmatrix} 0 \\ x^2 \end{pmatrix} + b_1 \begin{pmatrix} x^2 \\ xy \end{pmatrix} + a_2 \begin{pmatrix} 0 \\ x^3 \end{pmatrix} + b_2 \begin{pmatrix} 0 \\ x^2 y \end{pmatrix} + \sum_{k=3}^{\infty} [a_k \psi_k + b_k \chi_k] . \tag{53}$$

Moreover if $a_1 = 0$ and $a_2 \neq 0$, we can always reduce to $b_3 = 0$.

If we remove the non-degeneracy assumption $f_1 \neq 0$, most of the discussion and results remain essentially the same, and we have

Proposition 2. *If in the standard normal form expansion we have $f_k = 0$ for $k < m$, and f_m does not vanish, the system can always be written (formally) as*

$$\dot{\mathbf{x}} = A\mathbf{x} + a\psi_m + b\chi_m + c\chi_{m+1} + \sum_{k=m+2}^{\infty} [a_k \psi_k + b_k \chi_k] . \tag{54}$$

It should be noted that, as stated above, the result obtained here, in Propositions 1 and 2, is not a strong reduction with respect to the standard Poincaré normal form.

7. The Hamiltonian Case

The Poincaré theory of normal forms has a counterpart in the Hamiltonian case. In this case, one can deal directly with the Hamiltonian and its normal form, rather than with the Hamiltonian vector field that this generates.

It turns out that the procedure for the reduction of Poincaré NF presented here can also be applied, pretty much in the same way, in the case of Birkhoff NF. We will show this extension here without the details, since they are similar to those for the above discussion, provided the appropriate parallels are done. Lie-Birkhoff transformations and their use are discussed in e.g. [47,123,240].

We will consider Hamiltonians defined in a neighborhood of the origin in the phase space $R^{2\ell} \equiv R^n$, and having a critical point in the origin. Notice that, by adding an inessential constant $c = -H(0)$ to $H(x)$, we can always require that $H(0) = 0$. We will take coordinates (we write all the coordinate indices as lower ones, no matter if contravariant or covariant, to simplify the notation) $q_1, ..., q_\ell; p_1, ..., p_\ell$ in R^n. We also denote these by $x_1, ..., x_n$, where it is meant that $x_i = q_i$ for $i \leq \ell$, and $x_i = p_{i-\ell}$ for $i > \ell$.

We consider scalar functions on R^n, and denote by Σ the space of formal power series $s : R^n \to R$; we also denote by Σ_k the space of homogeneous scalar functions of degree $(k-2)$ on R^n.

We can then write any Hamiltonian in Σ which has a critical point in the origin, choosing $H(0) = 0$, as

$$H(q,p) = \sum_{k=0}^{\infty} H_k(q,p) \tag{55}$$

where $H_k \in \Sigma_k$.

We define then in Σ the familiar antisymmetric Poisson bracket $\{.,.\} : \Sigma \times \Sigma \to \Sigma$ as

$$\{F,G\} = \sum_{i=1}^{\ell} \left[\frac{\partial F}{\partial q_i} \frac{\partial G}{\partial p_i} - \frac{\partial F}{\partial p_i} \frac{\partial G}{\partial q_i} \right] . \tag{56}$$

With the above notation for Σ_k, we have

$$\{.,.\} \; : \; \Sigma_k \times \Sigma_m \to \Sigma_{k+m} \, . \tag{57}$$

To any $H \in \Sigma$ we associate a Hamiltonian vector field $X_H : R^n \to \mathrm{T}R^n$, defined by

$$X_H = \frac{\partial H}{\partial p_i} \frac{\partial}{\partial q_i} - \frac{\partial H}{\partial q_i} \frac{\partial}{\partial p_i} \tag{58}$$

so that $X_H(F) = \{F, H\}$. It can be easily checked that

$$\{F,G\} = K \iff [X_F, X_G] = -X_K \, . \tag{59}$$

To any $F \in \Sigma$ we can associate its ad-operator $\mathrm{ad}_F \equiv \{F,.\}$ and its Lie operator $\mathcal{P}_F$, defined as the time-one action of the flow under $(-X_F)$. This is also written as

$$\mathcal{P}_F = \exp[\mathrm{ad}_F] = \sum_{r=0}^{\infty} \frac{1}{r!} (\mathrm{ad}_F)^r \ . \tag{60}$$

We will now consider changes of coordinates given by the action of $\mathcal{P}_F$ for $F \in \Sigma$, similarly to what we have done in the Poincaré case. Notice that again, when we consider only first non-trivial order terms – i.e., the action on H_m if $F \in \Sigma_m$ – this gives simply

$$q_i \to q_i + \partial F/\partial p_i \quad , \quad p_i \to p_i - \partial F/\partial q_i \ . \tag{61}$$

The advantage of considering Lie operators rather than the usual near-identity changes of coordinates lies in the fact that, being defined in terms of Poisson brackets, these are guaranteed to generate canonical transformations at all orders. Moreover, we are guaranteed of the invertibility of some relevant operators [47].

Now, as can be checked by direct computation (see [47] and above), the effect of the change of coordinates given by $\mathcal{P}_F$ is given by

$$\begin{aligned} H \ \to \ \widetilde{H} &= \sum_{r=0}^{\infty}\sum_{r=0}^{\infty} \frac{1}{r!} (\mathrm{ad}_F)^r (H) = \\ &= H + \{F,H\} + \frac{1}{2}\{F,\{F,H\}\} + \frac{1}{6}\{F,\{F,\{F,H\}\}\} + \dots \ , \end{aligned} \tag{62}$$

which is formally analogous to (10). If we insert in this the expansion (55) for H and consider an $F \in \Sigma_k$, we get the exact analogue of (11), i.e.

$$\widetilde{H}_m = \sum_{r=0}^{[m/k]} \frac{1}{r!} (\mathrm{ad}_F)^r H_{m-rk} \ . \tag{63}$$

Then, to any series (55) we associate a series of homological operators $\mathcal{L}_k$ given by

$$\mathcal{L}_k = \{H_k, .\} \equiv \mathrm{ad}_{H_k} \tag{64}$$

which have the same formal properties of the $\mathcal{L}_k$ considered before.

The computation will now easily proceed along the same lines, and what's more, using the same formal properties as in the Poincaré case. Indeed, we could repeat word by word (with the same symbols – which now have a different definition) the discussion given above, and arrive at the same results. We will not bore the reader with this repetition, but only give the final result; in this it is meant that again $F^{(k)} = \bigcap_{s=0}^{k-1} \mathrm{Ker}(\mathcal{L}_k)$.

Definition 3. *We say that the Hamiltonian* $H = \sum_k H_k$ *is in* Birkhoff renormalized form *up to order* n *if* $H_k \in F_k^{(k)}$ *for all* $k \le n$.

Proposition 3. *Any Hamiltonian can be brought into Birkhoff renormalized form up to any desired order* n *by means of a formal series of Lie transforms.*

It is worth mentioning that we can also proceed by iterated Bikhoff normalizations, exactly as we proceeded by iterated Poincaré normalizations in Sect. 5 of this chapter, and arrive at the same results. We would then have explicitly:

Definition 4. *A polynomial Hamiltonian* $H = \sum_{k=0}^{\infty} H_k$ *is said to be in* n-th normal form *if* $H_k \in F_k^{(p)}$ *for all* k, *with* $p = \min(k,n)$. *When this condition is satisfied for all* k *(with no upper limit), we say that the system is in* Birkhoff renormalized form.

Notice that a Hamiltonian in n-th normal form is also in renormalized form up to order n, as was the case for dynamical systems (vector fields, formal power series).

Proposition 4. *Given a Hamiltonian* $H \in \Sigma$, $H = \sum_{k=0}^{\infty} H_k^{(0)}$, *it is always possible to reduce it, by a sequence of* n *formal Birkhoff normalizations, into the* n*-th normal form*

$$H = \sum_{k=0}^{n} H_k^{(k)} + \sum_{s=n+1}^{\infty} H_s^{(n)} ,$$

where $H_k^{(k)} \in F_k^{(k)}$ *for all* $k \le n$. *Formally, considering infinite sequences of Birkhoff normalizations, a Hamiltonian can always be taken into renormalized form.*

8. Renormalized Forms in the Presence of Symmetry

When we deal with standard Poincaré normal forms, it is well known [16,121,190] that symmetric systems exhibit symmetric normal forms.

More precisely, consider a one-parameter group G generated by the matrix S. The system $\dot{x} = f(x)$ is G-*equivariant* if

$$f(Gx) = (\mathrm{d}G) \cdot f(x) , \tag{65}$$

which in terms of S means simply

$$f^i(x + \varepsilon S x) \equiv f^i(x) + \varepsilon (\partial f^i / \partial x^j) S^j{}_k x^k = f^i(x) + \varepsilon S^i{}_j f^j(x) , \tag{66}$$

i.e. $Sx \cdot \nabla f = Sf$.

In this case, one can show [38,39,40,121,190] (see also the neat discussion in [169]) that it is possible to choose the functions h_k, generating the normalizing transformation, to be G-symmetric, and that the normal form

$$\dot{x} = \widetilde{f}(x) = \sum_{k=0}^{\infty} \widetilde{f}_k(x) \tag{67}$$

will also be G-equivariant. Introducing the operator

$$\mathcal{S} := \{Sx, .\} \tag{68}$$

this means that actually we have

$$\widetilde{f} \in \mathrm{Ker}(\mathcal{L}_0^+) \cap \mathrm{Ker}(\mathcal{S}) . \tag{69}$$

Remark 4. In other chapters, we have considered *nonlinear* symmetries, i.e. those corresponding to the action of a nonlinear vector field X_s such that $[X_f, X_s] = 0$ (where X_f corresponds to $\dot{x} = f(x)$). Here we will only consider linear symmetries.

Remark 5. The discussion in [169] also shows how to deal with the case where G does not coincide with its closure $\bar{G}$ in $\mathbf{R}^n$ (in this case we should consider $\bar{G}$ rather than G). Since elements of the form $e^{\lambda S}$ commute for different $\lambda \in \mathbf{R}$, G and $\bar{G}$ are abelian, and actually $\bar{G}$ is either $\mathbf{T}^m$ or $\mathbf{T}^m \times \mathbf{R}$ (with suitable m) [169,XVI.5].

In the case of the Poincaré renormalized forms that we are considering here, the same result holds, as we will discuss in this section. Here again we adopt the Bargmann scalar product.

First of all, we notice that as $f \in \mathrm{Ker}(\mathcal{S})$, we have in particular $f_0 \in \mathrm{Ker}(\mathcal{S})$ and thus $[A, S] = 0$.

Lemma. *If* $\mathcal{A} = \{Ax, .\}$ *and* $\mathcal{B} = \{Bx, .\}$, *then* $[\mathcal{A}, \mathcal{B}] = \{Cx, .\}$ *with* $C = -[A, B]$.

`Proof.` For any f, we have

$$[\mathcal{A}, \mathcal{B}]f = \{Ax, \{Bx, f\}\} - \{Bx, \{Ax, f\}\} \tag{70}$$

and using the Jacobi identity this reduces to $\{\{Ax, Bx\}, f\} \equiv \{Cx, f\}$, as is seen by elementary computations. ⊙

Corollary. *If* $[A, B] = 0$, *then* $[\mathcal{A}, \mathcal{B}] = 0$.

In our case, we are thus guaranteed that $\mathcal{L}_0$ and $\mathcal{S}$ commute. The same applies to $\mathcal{L}_0^+$ and $\mathcal{S}^+$. This also means, of course, that $\mathcal{S} : \mathrm{Ker}(\mathcal{L}_0) \to \mathrm{Ker}(\mathcal{L}_0)$ and conversely $\mathcal{L}_0 : \mathrm{Ker}(\mathcal{S}) \to \mathrm{Ker}(\mathcal{S})$. Similar formulas hold for $\mathcal{L}_0^+$ and $\mathcal{S}^+$. When A and/or S are non-normal matrices one can proceed along the lines of the discussion presented in Sects. 10 and 12 of Chap. 4 and, for the Hamiltonian case, Sects. 5 and 6 of Chap. 5.

Let us now suppose we perform the standard Poincaré normalization, i.e. we go to

$$\dot{x} = f^{(1)}(x) = \sum_{k=0}^{\infty} f_k^{(1)}(x) , \tag{71}$$

where not only $f^{(1)} \in \mathrm{Ker}(\mathcal{L}_0^+)$ but, thanks to the equivariant normal form theorem mentioned above [16,38,39,40,121,190], we are also guaranteed that

$$f_k^{(1)} \in \mathrm{Ker}(\mathcal{L}_0^+) \cap \mathrm{Ker}(\mathcal{S}) \cap V_k . \tag{72}$$

At the next stage, we will consider

$$f_k^{(2)} = f_k^{(1)} - \mathcal{L}_1 \left(h_{k-1}^{(1)} \right) \tag{73}$$

with $h^{(1)} \in \mathrm{Ker}(\mathcal{L}_0)$. Let us denote by $\mathcal{R}^{(1)}$ the range of $\mathcal{M}_1$, i.e. of $\mathcal{L}_1$ restricted to $\mathrm{Ker}(\mathcal{L}_0)$, and by $\mathcal{R}_s^{(1)}$ the range of $\mathcal{M}_1^s$, the restriction of $\mathcal{L}_1$ to $\mathrm{Ker}(\mathcal{L}_0) \cap \mathrm{Ker}(\mathcal{S})$.

By our standard procedure, $h_{k-1}^{(1)}$ is a solution to

$$h_{k-1}^{(1)} = \mathcal{M}_1^+ \left[P_1 f_k^{(1)} \right] . \tag{74}$$

We want to show that we can always choose $h^{(1)} \in \mathrm{Ker}(\mathcal{S})$. This is equivalent to showing that

$$\mathcal{R}^{(1)} \cap \mathrm{Ker}(\mathcal{S}) = \mathcal{R}_s^{(1)} . \tag{75}$$

Let us consider $f \in \mathcal{R}^{(1)} \cap \mathcal{S}$. We have that, for some $h \in \mathrm{Ker}(\mathcal{L}_0)$,

$$\begin{cases} f = \mathcal{L}_1(h) \\ \mathcal{S}(f) = 0 \end{cases} . \tag{76}$$

Applying $\mathcal{S}$ on the first of these, we have

$$\mathcal{S}\mathcal{L}_1(h) = 0 . \tag{77}$$

Let us now consider $[\mathcal{S}, \mathcal{L}_1]$. Since this corresponds – see the lemma above – to $\{Sx, f_1\}$, we are guaranteed by $f_1 \in \mathrm{Ker}(\mathcal{S})$ that (77) also means

$$\mathcal{L}_1\mathcal{S}(h) = 0 . \tag{78}$$

Note that this holds for any h such that $f = \mathcal{L}_1(h)$, and that such h are defined modulo $\mathrm{Ker}(\mathcal{L}_1)$. Since S generates a group and e^{Sx} is thus invertible, we can conclude that $h \in \mathrm{Ker}(\mathcal{S})$, or more precisely that if $h = h_0 + h_1$, with $h_1 \in \mathrm{Ker}(\mathcal{L}_1)$, then $h_0 \in \mathrm{Ker}(\mathcal{S})$ (note that $[\mathcal{L}_1, \mathcal{S}] = 0$ implies that $\mathrm{Ker}(\mathcal{L}_1)$ is invariant under $\mathcal{S}$).

We have thus shown that (75) is verified. It is clear that we can, exactly in the same way, prove that

$$\mathcal{R}^{(p)} \cap \mathrm{Ker}(\mathcal{S}) = \mathcal{R}_s^{(p)} \tag{79}$$

(with obvious definitions of $\mathcal{R}^{(p)}, \mathcal{R}_s^{(p)}$).

This also means that we have a complete extension of the results holding in the standard Poincaré equivariant case:

Proposition 5. *When transforming a system in Poincaré renormalized form, it is always possible to choose the generators $h_k^{(p)}$ of the normalizing transformations to be in* $\mathrm{Ker}(\mathcal{S})$*, so that the Poincaré renormalized form will commute with $X_S = (Sx)\partial_x$. In other words, it is always possible to obtain $f_k^{(k)} \in \mathrm{Ker}(\mathcal{S})$.*

With reference to Sect. 4 of this chapter, this means that in our functional setting we can intersect all the relevant spaces with (and restrict all relevant operators to) the space $\mathrm{Ker}(\mathcal{S})$.

Remark 6. As pointed out to us by Prof. Duistermaat some time ago, the results on equivariant standard normal forms can be derived at once by an argument based on filtration of Lie algebras [61,62,65]. The same argument also applies to the renormalized forms we are considering here, and thus it is no surprise that the equivariant settings are completely analogous.

Remark 7. The whole discussion conducted in this section on renormalized forms in the presence of symmetry would extend immediately to the Birkhoff-Gustavson normal forms, i.e. to the Hamiltonian framework considered above.

IX. Asymptotic Symmetries

We have been so far (except for one section) studying exact symmetries for differential equations. The point we would like to make in this final, short, chapter, is that in real applications one is often interested not so much in solutions with a given *exact symmetry*, but rather in solutions which are *asymptotically* (in time and/or space) symmetric.

As a simple example for the occurrence of such a situation, consider the ODE $dy/dt = -2ty^2$ which is invariant under the scaling $t \to \lambda t$, $y \to \lambda^{-2} y$ and has as a solution $y(t) = 1/(1+t^2)$. This solution is not invariant under the above scaling, but in the limit $t \to \pm\infty$ the scaling symmetry is asymptotically recovered.

We want to discuss here an approach to *asymptotic symmetry* of differential equations.

In doing this we will follow closely [147]. It should be stressed that this is actually just an outcome of an attempt to translate into the language of Lie theory the renormalization group (RG) approach pioneered by Goldenfeld and Barenblatt [28-30,165-167] and further developed by Bricmont and Kupiainen [58,59]. More precisely, we develop a kind of infinitesimal RG approach in the language and framework of Lie theory.

Several authors have considered approximate symmetries of differential equations [94,159] (these are also discussed in previous chapters), which can also be seen as asymptotic symmetries (for $|x| \to 0$) of ODEs,[1] but we are not aware, apart from these and the works quoted above, of any other attempt to study asymptotic symmetries. Moreover, as the above simple example shows, one can have asymptotic symmetries that are not symmetries and not even approximate symmetries out of the asymptotic region. In view of this remark, it will not be a surprise that this chapter is unrelated to the rest of the book. Also, we will find it convenient to discuss both ordinary and partial differential equations at the same time.

It should also be mentioned that this chapter is different from other ones in that, while the others report precise results and a pretty well formed theory, here we propose an approach that has not been so completely developed. Correspondingly, we will be less rigourous.

In the following, the total space M (to be defined below) will simply be a product of real spaces, i.e. will have a trivial topology. It would obviously be of

[1] We remark that approximate symmetries also play an important role in the study of perturbations of completely integrable evolution PDEs [133-135,205,343].

interest, both theoretically and for practical applications, to consider topologically nontrivial situations. However, we will not attempt such an analysis. This is motivated not only by a desire for simplicity, but also because the present approach is primarily intended to study physical phenomena for which the trivial topology of the total space is the natural one [147].

1. Notation and Basic Sets

Let us start by recalling some of the general notation introduced in Chap. I. We will denote by x the independent variables (at this stage we don't distinguish between space and time variables) and by u the dependent ones; x and u will belong, respectively, to spaces X and U, which will be embedded smooth submanifolds of $\mathbf{R}^q$ and $\mathbf{R}^p$. The *total space* will be

$$M = X \times U \ . \tag{1}$$

We consider vector fields on M, i.e. elements Y of $\mathcal{M} = \mathcal{D}\mathrm{iff}(M)$. These will be written as

$$Y = \xi^i(x,u)\frac{\partial}{\partial x^i} + \varphi^\alpha(x,u)\frac{\partial}{\partial u^\alpha} \ . \tag{2}$$

Such a vector field can be prolonged to a vector field $Y^{(n)}$ on $J^n M$, by the standard prolongation formula, see Chap. I.

As discussed above for ODEs (Chap. III), we will focus on vector fields which preserve the role of independent variables, i.e. those of the form

$$Y = \xi^i(x)\frac{\partial}{\partial x^i} + \varphi^\alpha(x,u)\frac{\partial}{\partial u^\alpha} \ . \tag{3}$$

These form the algebra

$$\mathcal{M}_G = \mathcal{D}\mathrm{iff}(X) \oplus_\rightarrow \mathcal{D}\mathrm{iff}(U) \tag{4}$$

which we will consider from now on. Here $\oplus_\rightarrow$ denotes the semidirect sum. Note that in the case of ODEs we thus have something more general than simply the Lie point time-independent symmetries.

A function $u = f(x)$ corresponds to a graph $\gamma_f = \{(x,u) \ : \ u = f(x)\} \subset M$; we can also consider a fiber bundle (M, X, π) with base X and projection $\pi : (x,u) \to x$; the vector fields in $\mathcal{M}_G$ preserve the fibered structure.

Thus, f corresponds to a section γ of this bundle. We will denote by Γ the set of smooth sections of the bundle. This is not a complete space. To avoid this problem, we should consider a Sobolev space of sections of (M, X, π), and its Sobolev completion (in suitable Sobolev norm). In the following, we will tacitly assume we have done so, and Γ will denote such a Sobolev completion, to which all the section considered will belong. Accordingly, we will consider a suitable Sobolev class of vector fields in $\mathcal{M}_G$, leaving $\mathcal{M}_G$ globally invariant.[2]

[2] In the following, the "Sobolev considerations" for vector fields, sections and their spaces, will be implicit, in order to avoid boring repetitions.

2. Induced Actions on Functions and Equations

Working formally, the action of $Y \in \mathcal{M}_G$ induces an action in the space Γ of smooth sections of (M, X, π). Under this a section $\gamma = \gamma_f$ will be transformed into a new section $\gamma' = \gamma_{f'}$. In other words Y induces an action in the space of smooth functions $f : X \to U$.

This action can be determined as follows: we have that points (x, u) are acted upon as

$$(I + \varepsilon Y)(x,u) = (\widetilde{x}, \widetilde{u}) = (x + \varepsilon\xi(x), u + \varepsilon\varphi(x,u)) \tag{5}$$

and we want to find an $\widetilde{f}$ such that $\widetilde{u} = \widetilde{f}(\widetilde{x})$. By the above formula, at first order in ε we have

$$\begin{aligned} \widetilde{u} =& f(x) + \varepsilon\varphi(x,u) = f\left(\widetilde{x} - \varepsilon\xi(x)\right) + \varepsilon\varphi\left(\widetilde{x} - \varepsilon\xi(x), u\right) = \\ =& f(\widetilde{x}) + \varepsilon\left[\varphi\left(\widetilde{x} + O(\varepsilon), f(\widetilde{x}) + O(\varepsilon)\right) - (Df)\cdot\xi\right] + O(\varepsilon^2) = \\ =& f(\widetilde{x}) + \varepsilon\left[\varphi\left(\widetilde{x}, f(\widetilde{x})\right) - (\xi \cdot Df)\right] \end{aligned} \tag{6}$$

so that the action of Y in Γ can be described by the formal vector field $\widetilde{Y} : \Gamma \to \mathrm{T}\Gamma$ which acts as

$$\begin{aligned} \left(e^{\varepsilon\widetilde{Y}} f\right)^\alpha (x) =& f^\alpha(x) + \varepsilon\delta f^\alpha(x) + O(\varepsilon^2) \\ (\widetilde{Y} f)^\alpha \equiv \delta f^\alpha(x) =& \varphi^\alpha\left(x, f(x)\right) - \xi^i \frac{\partial f^\alpha(x)}{\partial x^i} \ . \end{aligned} \tag{7}$$

Note that if we introduce the space $\Gamma_S \subset \Gamma$ of sections in Γ that are the graph of solutions to Δ (i.e. of the $\gamma \in \Gamma$ such that $\gamma \subset S_\Delta$), then for $Y \in \mathcal{G}_\Delta$ we have $\widetilde{Y} : \Gamma_S \to \mathrm{T}\Gamma_S$. This is indeed a direct translation of the statement that a symmetry of the equation transforms solutions into solutions.

In a similar way, the action of $Y^{(n)}$ in J^nM also transforms smooth submanifolds of J^nM (as the solution manifold S_Δ) into smooth submanifolds of J^nM. Since equations are identified with their solution manifolds, this also induces an action in the set E^n_M of equations of order n with variables in M.

Here we are mainly interested in *evolution equations*, in which case some extra care must be taken to take into account the special role of time with respect to the other (spatial) independent variables, which we now discuss.

We will denote by x the spatial independent variables, and by t the time variable. An evolution equation will be one of the form

$$\Delta :\equiv \quad u_t = G\left(x, t; u^{[n]}\right) \ , \tag{8}$$

where σ_Δ is a smooth function of its arguments and $u^{[n]}$ represents dependent variables and their partial derivatives *with respect to space variables only* up to order n.

In this setting, J^nM can be seen as a bundle whose fiber corresponds to the space of t-derivatives of the u's, and the base is the space of independent and dependent variables together with their derivatives of order up to n with respect to the space variables; this will be called the n-th *spatial jet*, J^n_xM.

With κ the projection $\kappa : J^nM \to J^n_xM$, the evolution equations in E^n_M are then in correspondence with elements of the space Σ of sections of (J^nM, J^n_xM, κ). In other words, the evolution equation (8) is identified with the section $\sigma_\Delta \in \Sigma$ given by the graph of G.

Note that in this setting the section σ_Δ corresponds to the solution manifold S_Δ: i.e., a point $(\alpha, \beta) \in J^nM$, where $\alpha \in J^n_xM$, $\beta \in \kappa^{-1}(\alpha)$, is in S_Δ if and only if $\beta = \sigma_\Delta(\alpha)$.

Note also that now the condition $Y \in \mathcal{M}_G$ is not sufficient to ensure that $Y^{(n)} : \Sigma \to \mathrm{T}\Sigma$. To this aim we must ask instead that $Y \in \mathcal{M}^t_G \subset \mathcal{M}_G$, where $\mathcal{M}^t_G$ is given by

$$\mathcal{M}^t_G = [\mathcal{D}\mathrm{iff}(T) \oplus_{\rightarrow} \mathcal{D}\mathrm{iff}(X)] \oplus_{\rightarrow} \mathcal{D}\mathrm{iff}(U) \ . \tag{9}$$

Here and in the following, $t \in T$, $x \in X$, and

$$M = (T \times X) \times U \ . \tag{10}$$

Elements of $\mathcal{M}^t_G$ will be written in the form

$$Y = \tau(t)\frac{\partial}{\partial t} + \xi^i(x,t)\frac{\partial}{\partial x^i} + \varphi^\alpha(x,t;u)\frac{\partial}{\partial u^\alpha} \ . \tag{11}$$

Now the action induced by Y in Σ can be determined pretty much in the same way as done for $\widetilde{Y}$ above. We write equations in the form of (8), and have that

$$\begin{aligned} \left(e^{\varepsilon \widehat{Y}} G\right)^\alpha &= G^\alpha + \varepsilon(\delta G)^\alpha + O(\varepsilon^2) \\ (\delta G)^\alpha &= (\tau_t G^\alpha - \varphi^\alpha_t) + \tau\frac{\partial G^\alpha}{\partial t} + \xi^i\frac{\partial G^\alpha}{\partial x^i} + \Phi^\beta_J \frac{\partial G^\alpha}{\partial u^\beta_J} \ , \end{aligned} \tag{12}$$

where we have used again the multiindex notation and the Φ^β_J are given by the prolongation formula.

Summarizing the discussion up to now, we have considered vector fields in $\mathcal{M}$, and then restricted our attention to $\mathcal{M}^t_G \subset \mathcal{M}$. For any vector field $Y \in \mathcal{M}^t_G$, we have considered associated vector fields $Y^{(n)}$, $\widetilde{Y}$, $\widehat{Y}$, acting in the spaces indicated in (13) below. Then, $Y \in \mathcal{G}_\Delta \Leftrightarrow \widetilde{Y} : \Gamma_S \to \mathrm{T}\Gamma_S$.

$$\begin{array}{ccccc} Y & : & M & \to & TM \\ Y^{(n)} & : & J^nM & \to & T(J^nM) \\ \widetilde{Y} & : & \Gamma & \to & T\Gamma \\ \widehat{Y} & : & \Sigma & \to & T\Sigma \end{array} \tag{13}$$

3. Symmetries and Asymptotic Symmetries

According to the standard definitions (Chap. I), the symmetry algebra $\mathcal{G}_\Delta$ of the equation $\Delta \in E^n_M$ is the algebra

$$\mathcal{G}_\Delta = \{ Y \in \mathcal{M} \ / \ Y^{(n)} : S_\Delta \to \mathrm{T} S_\Delta \} \subseteq \mathcal{M} \ . \tag{14}$$

Similarly, for a function $f : X \to U$, the symmetry algebra $\mathcal{G}_f$ is defined as

$$\mathcal{G}_f = \{ Y \in \mathcal{M} \ / \ Y : \gamma_f \to \mathrm{T}\gamma_f \} \ . \tag{15}$$

As discussed in Chap. III, in this context we actually want $Y \in \mathcal{M}_G \subset \mathcal{M}$ in (14,15).

We recall that for f a solution of Δ, i.e. such that $\gamma_f^{(n)} \subset S_\Delta$, it is not necessary that $\mathcal{G}_f \subseteq \mathcal{G}_\Delta$. The vector fields that are in $\mathcal{G}_f$ for some solution f but not in $\mathcal{G}_\Delta$ correspond to *conditional symmetries* [53,101,219,272,328,329,341,342].

Physically, conditional symmetries correspond to the possibility of solutions having symmetries that are *not* symmetries of the equation. To give a simple example, consider the equation

$$\left(\partial_x^2 + \partial_y^2\right) f = -2\alpha f + \alpha^2 (x^2 + y^2) f + \beta y^2 f + (\beta/\alpha) y \partial_y f \tag{16}$$

($\alpha \neq 0, \beta$ real constants) which is not symmetric under rotations in the (x, y) plane, and admits the solution

$$f(x, y) = e^{-\alpha(x^2+y^2)/2} \tag{17}$$

which is rotationally symmetric. See e.g. [341,342] for other examples, less simple and relevant to applications.

For later discussion, we express the symmetry condition for equations and solutions in terms of the $\widetilde{Y}$ and $\widehat{Y}$ introduced above. It is clear that

$$\mathcal{G}_\Delta = \{ Y \ / \ e^{\lambda \widehat{Y}} \sigma \Delta = \sigma \Delta \quad \forall \lambda \in \mathbf{R} \} \tag{18}$$

$$\mathcal{G}_f = \{ Y \ / \ e^{\lambda \widetilde{Y}} \gamma_f = \gamma_f \quad \forall \lambda \in \mathbf{R} \} \tag{19}$$

are equivalent to our previous definitions (14,15).

Let us look again at this from a slightly different point of view. We consider $\widetilde{Y}$ and $\widehat{Y}$ as defining a (formal) dynamical system in the spaces Γ and Σ, which we write as

$$\frac{\mathrm{d}\gamma}{\mathrm{d}\lambda} = \widetilde{Y} \gamma \tag{20}$$

$$\frac{\mathrm{d}\sigma}{\mathrm{d}\lambda} = \widehat{Y} \sigma \ . \tag{21}$$

Here the evolution is with respect to an evolution parameter λ [see (18,19)] which is *not* the physical time.

We write formally the solutions to the evolution equations (20,21) as

$$\Phi(\lambda,\gamma_0) = e^{\lambda\widetilde{Y}}\gamma_0 \tag{22}$$
$$\Psi(\lambda,\sigma_0) = e^{\lambda\widehat{Y}}\sigma_0 \ , \tag{23}$$

where we have used a notation that is standard in dynamical systems theory: i.e. if $\gamma(\lambda_0) = \gamma_0$, then $\gamma(\lambda) = \Phi(\lambda - \lambda_0, \gamma_0)$, and the like for $\sigma(\lambda)$. We can then restate our symmetry criterion as follows:

Symmetry criterion for functions. *A function f is symmetric under Y if and only if the corresponding section $\gamma_f \in \Gamma$ is a fixed point for the flow generated by $\widetilde{Y}$.*

Symmetry criterion for equations. *An equation Δ is symmetric under Y if and only if the corresponding section $\sigma_\Delta \in \Sigma$ is a fixed point for the flow generated by $\widehat{Y}$.*

The asymptotic symmetries will be defined by means of this formulation. In order to do this, we have to consider the flows (20,21) in greater detail.

In particular, we should focus on functions (respectively, equations) which are not a fixed point for $\widetilde{Y}$ (respectively, for $\widehat{Y}$), but which go to a fixed point[3] under the evolution (20), i.e. under the flow of $\widetilde{Y}$ [respectively, under the evolution (21), i.e. under the flow of $\widehat{Y}$]. We will thus define asymptotic symmetries[4] as follows:

Definition 1. *The vector field $Y \in \mathcal{M}_G$ is an ω_λ-*asymptotic symmetry *of the function $f : X \to U$ if and only if*

$$\lim_{\lambda\to\infty} \Phi(\lambda,\gamma_f) = \gamma_f^\infty$$

*is a fixed point under $\widetilde{Y}$, i.e. $e^{\lambda\widetilde{Y}}\gamma_f^\infty = \gamma_f^\infty$ for all λ. The vector field $Y \in \mathcal{M}_G$ is an α_λ-*asymptotic symmetry *of the function $f : X \to U$ if and only if*

$$\lim_{\lambda\to-\infty} \Phi(\lambda,\gamma_f) \equiv \lim_{\lambda\to\infty} \Phi^*(\lambda,\gamma_f) = \gamma_f^{-\infty}$$

is a fixed point under $\widetilde{Y}$, i.e. $e^{\lambda\widetilde{Y}}\gamma_f^{-\infty} = \gamma_f^{-\infty}$ for all λ.

Notice that, since Y is a diffeomorphism, it is legitimate to consider time-reversed flows, and γ_0 is a fixed point under $\widetilde{Y}$ if and only if it is a fixed point under $-\widetilde{Y}$. Notice also that fixed points under $\widetilde{Y}$ are the functions f such that $\widetilde{Y}(\gamma_f) = 0$.

We can also give an analogous definition for the asymptotic symmetries of equations:

Definition 2. *The vector field $Y \in \mathcal{M}$ is an ω_λ-*asymptotic symmetry *of the equation $\Delta \in E^n_M$ if and only if*

---

[3] These are obviously on stable manifolds (unstable ones when considering the reversed flow) of the fixed point in question, see Chap. II and references therein.

[4] As customary in dynamical systems literature, we denote by ω the limit obtained for "forward" evolution, and by α the limits obtained for "backward" evolution.

$$\lim_{\lambda\to\infty} \Psi(\lambda, \sigma_\Delta) = \sigma_\Delta^\infty$$

is a fixed point under $\widehat{Y}$*, i.e.* $e^{\lambda\widehat{Y}}\sigma_\Delta^\infty = \sigma_\Delta^\infty$ *for all* λ*. The vector field* $Y \in \mathcal{M}$ *is an* α_λ-asymptotic symmetry *of the equation* $\Delta \in E^n_M$ *if and only if*

$$\lim_{\lambda\to-\infty} \Psi(\lambda, \sigma_\Delta) \equiv \lim_{\lambda\to\infty} \Psi^*(\lambda, \sigma_\Delta) = \sigma_\Delta^{-\infty}$$

is a fixed point under $\widehat{Y}$*, i.e.* $e^{\lambda\widehat{Y}}\sigma_\Delta^{-\infty} = \sigma_\Delta^{-\infty}$ *for all* λ*.*

It is quite clear that the name "asymptotic symmetry" is justified: if Y is an ω_λ-asymptotic symmetry of f, then under the flow of $\widetilde{Y}$ the function f is driven to a function f_0 for which Y is an ordinary symmetry, and similarly for σ_Δ, and/or for α_λ-asymptotic symmetries.

We would like to emphasize again that "asymptotic" should be understood with respect to λ. I.e., it refers to the flows induced by the vector field Y. Asymptotic properties in this sense can correspond or not correspond to asymptotic properties in time and/or space, depending on the Y being considered. This should be kept in mind for the following discussion, in order to avoid confusion.

4. Asymptotic Symmetries and Space-Time Asymptotic Properties

In the previous section, we have provided a notion of asymptotic symmetry. We would like to now discuss briefly how this can be used in practice.

As we have seen above, when studying a differential equation Δ_0, we can consider a vector field Y which is or is not a symmetry for Δ_0. We will consider the two cases separately. For ease of notation, we will denote the entire set of independent variables (i.e. both spatial and time ones) by x, and by X their space.

Let us first assume that Y is an exact symmetry of Δ_0, that is, in our present language, that

$$\widehat{Y}\Delta_0 = 0 \, . \tag{24}$$

In this case the solution manifold $S_\Delta \subset J^n M$ is invariant under the flow of $Y^{(n)}$. The space Γ_S of sections that are the graph of solutions to Δ_0 is therefore also invariant under $\widetilde{Y}$. Namely, (24) implies that

$$Y^{(n)} : S_\Delta \to \mathrm{T}S_\Delta \quad ; \quad \widetilde{Y} : \Gamma_S \to \mathrm{T}\Gamma_S \, . \tag{25}$$

Let us assume now that $\lim[e^{\lambda Y}x; \lambda \to \pm\infty]$ describes the spatio-temporal asymptotic region we are interested in (e.g. $|x| \to \infty$ and/or $t \to \infty$), for all $x \in X$. Then, the asymptotic behavior (in this spatio-temporal region) of a solution $u = f(x)$ to Δ_0 is described by the limit $f_{\pm\infty}$ of $e^{\lambda\widetilde{Y}}f$ for $\lambda \to \pm\infty$.

In particular, if the flow of $\widetilde{Y}$ in $\Gamma_S \subset \Gamma$ happens to be ω_λ−asymptotic (respectively, α_λ−asymptotic) to a lower dimensional set $\Gamma_S^{(\omega)}$ (respectively $\Gamma_S^{(\alpha)}$, the elements of this set give a classification of the asymptotic behavior of solutions to Δ_0. In the renormalization group language, we say that they identify *universality classes.*

We have therefore a setting (including motivation and use) very similar to the one introduced in the Bricmont-Kupiainen analysis [58,59] analysis of the discrete renormalization group approach to parabolic PDEs and to the asymptotic scaling behavior of solutions. The present setting extends their idea to the infinitesimal renormalization group, and to general (provided it can be described in terms of Lie groups) asymptotic behavior.

Let us now consider the other case, i.e., we consider a vector field Y that is *not* a symmetry of Δ_0. Thus, we now assume

$$\widehat{Y} \Delta_0 \neq 0 \, . \tag{26}$$

In this case, $\widehat{Y}$ induces a non-trivial flow in the space of equations. We will write

$$e^{\lambda \widehat{Y}} \Delta_0 = \Delta_\lambda \tag{27}$$

and will denote by $S_{\Delta_\lambda} \equiv S_\lambda$ the solution manifold of Δ_λ. As remarked above, Δ_λ, S_λ and the associated section $\sigma_{\Delta_\lambda} \equiv \sigma_\lambda$ are just realizations of the same geometrical object in different languages, and thus (27) implies

$$e^{\lambda Y^{(n)}} S_0 = S_\lambda \, . \tag{28}$$

Let us now consider a solution f_0 to Δ_0, and denote by $\gamma_{f_0} \equiv \gamma_0$ its graph. We consider the flow originating from f_0 (that is, γ_0) under $\widetilde{Y}$, write

$$e^{\lambda \widetilde{Y}} f_0 = f_\lambda \, , \tag{29}$$

and denote as $\gamma_{f_\lambda} \equiv \gamma_\lambda$ the graph of f_λ. By (28),

$$\gamma_0 \subset S_0 \Longrightarrow \gamma_\lambda \subset S_\lambda \, , \tag{30}$$

so that while transforming the equation, Y also correspondingly transforms its solutions.

By letting λ go to $\pm\infty$ in (27-29), and writing $\Delta_{\pm\infty}$, $f_{\pm\infty}$ for the limits (we will assume they exist), we have:

Lemma 1. *If f_0 is a solution to Δ_0 and the limits $f_{\pm\infty}$, $\Delta_{\pm\infty}$ exist, then $f_{\pm\infty}$ is a solution to $\Delta_{\pm\infty}$.*

Corollary. *If $\Delta_0 \to \Delta_*$ for $\lambda \to \pm\infty$ under (27), then all the solutions f_0 of Δ_0 go to solutions f_* of Δ_* under (29).*

This permits again to classify equations in universality classes (identified by the limit equation Δ_*), and inside each class of equation we can give a classification of solutions in universality classes according to their limit behavior.

It should be stressed that if f_0 is a generic function such that f_* is a solution to Δ_*, and Δ_0 is taken to this Δ_* in the limit (with the above notation for these),

f_0 is *not* necessarily a solution to Δ_0. Namely, the fact that f_* is a solution to Δ_* is a necessary but not a sufficient condition for f_0 to solve Δ_0.

It should also be stressed that the classifications considered above are only able to capture those equations and functions whose asymptotic behavior under the flow induced by the considered vector field Y is particularly simple, indeed a fixed point. This behavior is by no means a trivial feature, and actually for a generic Y we will have no solutions admitting such a limit. This just means that no solution have the asymptotic symmetry described by the Y we are considering.

It would be natural to ask how one could determine *a priori* the possible asymptotic behaviors within the approach proposed here. We are not able to answer this question at present.[5]

Thus, as the method is based on testing a given asymptotic symmetry, this should be chosen on the basis of other considerations, e.g. its physical interest and/or the results of experiments in systems modelled by the equation under study.

On the whole, what is proposed here is a framework of discussing general asymptotic properties of (solutions to) differential equations, and not a complete theory. It is hoped that such a theory will be devised in the future.

Needless to say, our discussion has been purely formal. The examples to be considered in the exercises below (taken from [147]) will show that the framework developed here is not empty, and applies to physically meaningful equations.[6]

Example 1. Let us consider an autonomous dynamical system in R^n, so that $x \in X = R$, and $u \in R^n$. We write the equation as

$$\frac{\mathrm{d}u^i}{\mathrm{d}x} = F^i(u) \ . \tag{31}$$

We consider the vector field $Y = -\partial_x$, which is a symmetry of (31). The flow of $f : R \to R^n$ under $\widetilde{Y}$ is described by

$$\frac{\mathrm{d}f^i}{\mathrm{d}\lambda} = F^i(f) \ . \tag{32}$$

Thus, ω_λ-asymptotic symmetry under $\widetilde{Y}$ and x-asymptotic invariance under ∂_x are equivalent, and only solutions $u^i = f^i(x)$ such that $\lim_{x\to\infty} f(x)$ is a zero of F have ∂_x as ω_λ-asymptotic symmetry. Similarly, only solutions $u = f(x)$ such that $\lim_{x\to-\infty} f(x)$ is a zero of F have ∂_x as α_λ-asymptotic symmetry. Notice that solutions $f(x)$ having Y as both α_λ- and ω_λ-asymptotic symmetries

---

[5] It is an obvious consideration that all the Y satisfying (24) are good candidates and should be considered. This is not a fast procedure, but it is actually the same thing one does when one looks for special solutions to a given PDE by considering exact symmetries and the corresponding solutions.

[6] These exercises will deal only with dynamical systems. The reader wishing to consider examples of applications of the present scheme to PDEs is referred again to [147].

are either stationary ones, or homoclinic or heteroclinic orbits connecting fixed points. ⊙

Exercise 1. Consider the case where both the dependent and independent variable are in R, and consider the equation

$$\Delta_0 \equiv \frac{du}{dx} + cu = 0 ,$$

with c a positive real constant. Consider the vector field $Y = -\partial_x$, and check that this is a symmetry of Δ_0. Check that among the solutions $f_b = b \cdot e^{-cx}$, f_0 is the only fixed point for $\widetilde{Y}$, and that

$$\lim_{\lambda \to \infty} e^{\lambda \widetilde{Y}} f_b = f_0$$

for all $b \in R$. Compare this with what you know about solutions to Δ_0.

Exercise 2. Denote by $t \in R$ the independent variable, and by $(x, y) \in R^2$ the dependent ones; consider the dynamical system

$$\begin{aligned} \dot{x} &= +x - Ay - (x^2 + y^2)x \\ \dot{y} &= +y + Ax - (x^2 + y^2)y , \end{aligned}$$

where A is a real constant, and consider the two-parameter family of vector fields

$$Y_{(\alpha,\beta)} = \alpha \partial_t + \beta (x \partial_y - y \partial_x) ,$$

which are symmetries of the dynamical system. Denote a solution as $f = (x_f(t), y_f(t))$. Determine the equation describing the action of $\widetilde{Y}$ in Γ, and check that there is no non-trivial fixed point under $\widetilde{Y}$, unless $\beta = A\alpha$. Check also that $Y_{-1,-A}$ is an ω_λ-asymptotic symmetry, and $Y_{1,A}$ an α_λ-asymptotic symmetry, for all solutions. Interpret this in terms of asymptotic properties of f for $t \to \infty$ and for $t \to -\infty$, and check that indeed solutions of the dynamical system under consideration behave in such a way.

Exercise 3. Extend the analysis of the previous exercise to the more general case of

$$\begin{aligned} \dot{x} &= f(r)x - \Omega(r)y \\ \dot{y} &= f(r)y + \Omega(r)x \end{aligned}$$

where $r = \sqrt{x^2 + y^2}$, and $f(r)$, $\Omega(r)$ are smooth functions.

References

1. S. Abbate, D. Ghisletti, A. Giorgilli, L. Lespade, G. Longhi, "Characterization of vibrational transition modes by use of normal forms", *Theor. Chimica Acta* **87** (1993), 215-232
2. R. Abraham and J.E. Marsden, *Foundations of mechanics*, Benjamin, New York 1978
3. V.S. Afrajmovich, V.I. Arnold, Yu.S. Il'yashenko and L.P. Shil'nikov: *Theory of bifurcations*; in Encyclopaedia of Mathematical Sciences vol. 5 – Dynamical Systems V (V.I. Arnold ed.), Springer, Berlin 1989
4. D.V. Aleekseevskij, A.M. Vinogradov and V.V. Lychagin, *Basic ideas and concepts of differential geometry*, in Encyclopaedia of Mathematical Sciences vol. 28 – Geometry I (R.V. Gamkrelidze ed.), Springer, Berlin 1991
5. A. Algaba, E. Freire and E. Gamero, "Isochronicity via normal form", preprint 1998
6. M.K. Ali, "Quantization of the double pendulum", *Can. J. Phys.*, **74** (1996), 255-262
7. U. Amaldi and T. Levi Civita, *Lezioni di Meccanica Razionale*, Zanichelli, Bologna 1924
8. M. Andreolli, D. Bambusi and A. Giorgilli, "On a weakened form of the averaging principle in multifrequency systems", *Nonlinearity* **8** (1995), 283-293
9. D. Arnal, M. Ben Ammar and G. Pinczon, "The Poincaré–Dulac theorem for nonlinear representations of nilpotent Lie algebras", *Lett. Math. Phys.* **8** (1984), 467-476
10. L. Arnold and P. Imkeller, "Normal forms for stochastic differential equations", *Probability Theory and Related Fields"* **110** (1998), 559-588
11. V.I. Arnold, *Equations differentielles ordinaires*, M.I.R., Moscow, 1974, 2nd ed. 1990; V.I. Arnold: *Ordinary Differential Equations*, Springer, Berlin 1992
12. V. Arnold, *Chapitres supplementaires de la théorie des equations differentielles ordinaires*, M.I.R., Moscow, 1980; V.I. Arnold, *Geometrical methods in the theory of ordinary differential equations*, Springer, Berlin 1983
13. V. Arnold, *Les méthodes mathématiques de la mecanique classique*, M.I.R., Moscow, 1976; V.I. Arnold, *Mathematical methods of classical Mechanics*, Springer, Berlin 1983, 1989
14. V.I. Arnold, *Catastrophe theory*; in Encyclopaedia of Mathematical Sciences vol. 5 – Dynamical Systems V, (V.I. Arnold ed.), Springer, Berlin 1994
15. V.I. Arnold and A.B. Givental, *Symplectic Geometry*, in Encyclopaedia of Mathematical Sciences vol. 4 – Dynamical Systems IV (V.I. Arnold and S.P. Novikov eds.), Springer, Berlin 1988
16. V.I. Arnold and Yu.S. Il'yashenko, *Ordinary differential equations*; in: Encyclopaedia of Mathematical Sciences vol. 1 – Dynamical Systems I, (D.V. Anosov and V.I. Arnold eds.), pp. 1-148, Springer, Berlin 1988

17. V.I. Arnold, V.V. Kozlov and A.I. Neishtadt: *Mathematical aspects of classical and celestial mechanics*; in: Encyclopaedia of Mathematical Sciences vol. 3 – Dynamical Systems III, (V.I. Arnold ed.), 2nd edition, pp. 1-291, Springer, Berlin 1993
18. A. Baider, "Unique normal forms for vector fields and hamiltonians", *J. Diff. Eqs.* **78** (1989), 33
19. A. Baider and R.C. Churchill, "Uniqueness and non-uniqueness of normal forms for vector fields", *Proc. R. Soc. Edinburgh A* **108** (1988), 27
20. D. Bambusi, "A Nekhoroshev-type theorem for the Pauli-Fierz model of classical electrodynamics", *Ann. I.H.P. (Phys. Theo.)* **60** (1994), 340-371
21. D. Bambusi, "Exponential stability of breathers in Hamiltonian networks of weakly coupled oscillators", *Nonlinearity* **9** (1996), 433-457
22. D. Bambusi, "Long time stability of some small amplitude solutions in nonlinear Schroedinger equations", *Comm. Math. Phys.* **189** (1997) 205-226
23. D. Bambusi, "Nekhoroshev theorem for small amplitude solutions in nonlinear Schroedinger equation", *Math. Z.*, to appear (1999)
24. D. Bambusi, G. Cicogna, G. Gaeta, and G. Marmo: "Normal forms, symmetry, and linearization of dynamical systems", *Journ. Phys. A: Math. Gen.*, **31** (1998) 5065-5082
25. D. Bambusi and G. Gaeta eds., *Symmetry and Perturbation Theory* (proceedings Torino workshop 1996), C.N.R. 1998
26. D. Bambusi and A. Giorgilli, "Exponential stability of states close to resonance in infinite-dimensional Hamiltonian systems", *J. Stat. Phys.* **71** (1993), 569-606
27. D. Bambusi and N.N. Nekhoroshev, "A property of exponential stability in nonlinear wave equations near the fundamental linear mode", *Physica D* **122** (1998), 73-104
28. G.I. Barenblatt, *Dimensional analysis*, Gordon and Breach, New York 1987
29. G.I. Barenblatt, "Intermediate asymptotics, scaling laws and renormalization group in continuum mechanics", *Meccanica* **28** (1993), 177-183
30. G.I. Barenblatt, *Scaling, self-similarity, and intermediate asymptotics*, Cambridge University Press, Cambridge 1996
31. V. Bargmann, "On a Hilbert space of analytic functions and an associated integral transform", *Comm. Pure Appl. Math.* **14** (1961), 187
32. A.O. Barut and J.R.R. Zeni, *Group theory and differential equations*, World Scientific, Singapore 1999
33. D. Battig, T. Kappeler and B. Mityagin, "On the Korteweg-de Vries equation – convergent Birkhoff normal form", *J. Funct. Anal.* **141** (1996), 336-358
34. G. Baumann, *Symmetry analysis of differential equations with Mathematica*, Springer, Berlin 1999
35. A. Bazzani, G. Turchetti, P. Mazzanti, G. Servizi, "Normal forms for Hamiltonian maps and nonlinear effects in particle accelerators", *Nuovo Cimento B* **102** (1988), 51-80
36. A. Bazzani, M. Malavasi, S. Siboni, C. Pellacani, S. Rambaldi, G. Turchetti, "Poincaré map and anomalous transport in a magnetic confined plasma", *Nuovo Cimento B* **103** (1989), 659-668
37. A. Bazzani, E. Todesco, G. Turchetti, G. Servizi, "A normal form approach to the theory of nonlinear betatronic motion", CERN Report 94-02, Geneva 1994
38. G.R. Belitsky, "Equivalence and normal forms of germs of smooth mappings", *Russ. Math. Surveys* **33** (1978), 107

39. G.R. Belitsky, "Normal forms relative to the filtering action of a group", *Trans. Moscow Math. Soc.* **40** (1981), 1
40. G.R. Belitsky, "Smooth equivalence of germs of vector fields with a single eigenvalue or a pair of purely imaginary eigenvalues", *Funct. Anal. Appl.* **20** (1987), 253-259
41. G. Benettin, A. Carati and F. Fassò, "On the conservation of adiabatic invariants for a system of coupled rotators", *Physica D* **104** (1997), 253-268
42. G. Benettin, A. Carati and G. Gallavotti, "A rigorous implementation of the Jeans-Teller-Landau approximation for adiabatic invariants", *Nonlinearity* **10** (1996), 479-505
43. G. Benettin and F. Fassò, "Fast rotation of the rigid body: a study by Hamiltonian perturbation theory", *Nonlinearity* **9** (1996), 137-186
44. G. Benettin and F. Fassò, "From Hamiltonian perturbation theory to symplectic integrators and back" *Appl. Num. Math.* **29** (1998), 73-87
45. G. Benettin, F. Fassò and M. Guzzo, "Fast rotation of the rigid body: a study by Hamiltonian perturbation theory. Part II: gyroscopic rotations", *Nonlinearity* **10** (1997), 1695-1717
46. G. Benettin, J. Fröhlich and A. Giorgilli, "A Nekhoroshev-type theorem for Hamiltonian systems with infinitely many degrees of freedom", *Comm. Math. Phys.* **119** (1988), 95-108
47. G. Benettin, L. Galgani and A. Giorgilli, "A proof of the Kolmogorov theorem on invariant tori using canonical transformations defined by the Lie method", *Nuovo Cimento B* **79** (1984), 201
48. G. Benettin, L. Galgani and A. Giorgilli, "A proof of Nekhoroshev's theorem for the stability times in nearly integrable Hamiltonian systems", *Cel. Mech.* **37** (1985), 1-25
49. G. Benettin, L. Galgani and A. Giorgilli, "Realization of holonomic constraints and freezing of high frequency degrees of freedom in the light of classical perturbation theory I", *Comm. Math. Phys.* **113** (1987), 87-103
50. G. Benettin, L. Galgani and A. Giorgilli, "Realization of holonomic constraints and freezing of high frequency degrees of freedom in the light of classical perturbation theory II", *Comm. Math. Phys.* **121** (1989), 557-601
51. P. Bergé, Y. Pomeau, Ch. Vidal, *L'ordre dans le chaos*, Hermann, Paris 1984
52. G. Birkhoff, *Hydrodynamics*, Princeton University Press, 1960
53. G.W. Bluman and J.D. Cole, "The general similarity solution of the heat equation", *J. Math. Mech.* **18** (1969), 1025
54. G.W. Bluman and J.D. Cole, *Similarity methods for differential equations*, Springer, Berlin 1974
55. G.W. Bluman and S. Kumei, *Symmetries and differential equations*, Springer, 1989
56. N.N. Bogoliubov and V.A. Mitropolsky, *Asymptotic methods in the theory of nonlinear oscillations*, Hindustan, New Delhi 1961; *Méthodes asymptothiques dans la théorie des oscillations non-linéaires*, Gauthier-Villars, Paris 1962
57. J. Bourgain, "Construction of approximative and almost periodic solutions of perturbed linear Schrödinger and wave equations", *GAFA* **6** (1996), 201-230
58. J. Bricmont and A. Kupiainen, "Renormalization group and the Ginzburg-Landau equation", *Comm. Math. Phys.* **150** (1992), 193-208
59. J. Bricmont and A. Kupiainen, "Renormalizing Partial Differential Equations", in *Constructive Physics*, V. Rivasseau ed. (proceedings Ecole Polytechnique conference, 1994) Springer, Berlin 1995

60. T. Bridges, "Toral-equivariant partial differential equations and quasiperiodic patterns", *Nonlinearity* **11** (1998), 467-500
61. H.W. Broer, *Bifurcations of singularities in volume preserving vector fields*, Ph.D. Thesis, Groningen 1979
62. H.W. Broer, "Formal normal form theorems for vector fields and some consequences for bifurcations in the volume preserving case", in "Dynamical systems and turbulence", D.A. Rand and L.S. Young eds., it Lect. Notes Math. **898**, Springer, Berlin 1981
63. H.W. Broer, "Conservative dynamical systems", in *Structures in Dynamics* (H.W. Broer, F. Dumortier, S. van Strien and F. Takens eds.), North Holland, Amsterdam 1991
64. H.W. Broer, G.B. Huitema and M.B. Sevryuk, "Quasi periodic motions in families of dynamical systems: order amidst chaos", *Lect. Notes Math.* **1645**, Springer, Berlin 1996
65. H.W. Broer and F. Takens, "Formally symmetric normal forms and genericity", *Dynamics Reported* **2** (1989), 39-59
66. A.D. Bruno, "Analytical form of differential equations I", *Trans. Moscow Math. Soc.* **25** (1971), 131-288
67. A.D. Bruno, "Analytical form of differential equations II", *Trans. Moscow Math. Soc.* **26** (1971), 199-239
68. A.D. Bruno, "The normal form of a Hamiltonian system", *Russ. Math. Sur.*, **43** (1988) 25-66
69. A.D. Bruno, *Local Methods in the Theory of Differential Equations*, Springer, Berlin 1989
70. A.D. Bruno, "Divergence of the real normalizing transformation", *Selecta Math. (formerly Sovietica)* **12** (1993), 13-23
71. A.D. Bruno, "Ways of computing a normal form", *Dokl. Ak. Nauk*, **345** (1995) 298-300
72. A.D. Bruno, "Normal forms", *Math. and Computers in Simulation* **45** (1998), 413-427
73. A.D. Bruno and S. Walcher, "Symmetries and convergence of normalizing transformations", *J. Math. Anal. Appl.* **183** (1994), 571-576
74. J. Cariñena, M.A. Del Olmo and P. Winternitz, "On the relation between weak and strong invariance of differential equations", *Lett. Math. Phys.* **29** (1993), 151-163
75. J. Carr, R.G. Muncaster, "The application of center manifolds to amplitude expansions", *J. Diff. Eq.* **50** (1983), 260-279 and 280-288
76. A. Celletti and L. Chierchia, "On the stability of realistic three-body problems", *Comm. Math. Phys.* **186** (1997), 413-449
77. A. Celletti and L. Ferrara, "An application of the Nekhoroshev theorem to the restricted three-body problem", *Cel. Mech.* **64** (1996), 261-272
78. B. Champagne, W. Heremann and P. Winternitz, "The computer calculation of Lie-point symmetries for large systems of differential equations", *Comp. Phys. Comm.* **66** (1991), 319-341
79. L. Chierchia and C. Falcolini, "Compensations in small divisor problems", *Comm. Math. Phys.* **175** (1996), 135-160
80. P. Chossat and R. Lauterbach, *Methods in equivariant bifurcations and dynamical systems with applications*, World Scientific, Singapore 1999
81. S.N. Chow and J.K. Hale, *Methods of bifurcation theory*, Springer, Berlin 1982

82. S.N. Chow, C. Li and D. Wang, *Normal forms and bifurcations of planar vector fields*, Cambridge University Press, Cambridge 1994
83. L.O. Chua and H. Kokubu, "Normal forms for nonlinear vector fields. Part I: theory", *IEEE Trans. Circ. Syst.* **35** (1988), 863-880
84. L.O. Chua and H. Kokubu, "Normal forms for nonlinear vector fields. Part II: applications", *IEEE Trans. Circ. Syst.* **36** (1989), 851-70
85. R.C. Churchill and M. Kummer, "A unified approach to normal forms", *J. Symbol. Comp.* **27** (1999), 49-131
86. R.C. Churchill, M. Kummer and D.L. Rod, "On averaging, reduction and symmetry in hamiltonian systems", *J. Diff. Eqs.* **49** (1983), 359-414
87. G. Cicogna, "Symmetries of dynamical systems and convergent normal forms", *J. Phys. A* **28** (1995), L179-L182
88. G. Cicogna, "On the convergence of the normalizing transformation in the presence of symmetries", *J. Math. Anal. Appl.* **199** (1996), 243-255
89. G. Cicogna, "Convergent normal forms of symmetric dynamical systems", *J. Phys. A* **30** (1997), 6021-6028
90. G. Cicogna, "Multiple-periodic bifurcation and resonance in dynamical systems", *Nuovo Cimento B* **113** (1998), 1425-1430
91. G. Cicogna and G. Gaeta, "Periodic solutions from quaternionic bifurcation", *Lett. Nuovo Cim.* **44** (1985), 65-68
92. G. Cicogna and G. Gaeta, "Quaternionic-like bifurcation in the absence of symmetry", *J. Phys. A* **20** (1987), 79-89
93. G. Cicogna and G. Gaeta, "Nonlinear Lie symmetries in bifurcation theory", *Phys. Lett. A* **172** (1993), 361-364
94. G. Cicogna and G. Gaeta, "Approximate symmetries in dynamical systems", *Nuovo Cim. B* **109** (1994), 989-1008
95. G. Cicogna and G. Gaeta, "Poincaré normal forms and Lie point symmetries", *J. Phys. A* **27** (1994), 461-476
96. G. Cicogna and G. Gaeta, "Normal forms and nonlinear symmetries", *J. Phys. A* **27** (1994), 7115-7124
97. G. Cicogna and G. Gaeta, "Symmetry invariance and center manifolds in dynamical systems", *Nuovo Cim.* **B 109** (1994), 59-76
98. G.Cicogna and G.Gaeta: "Lie point symmetries and nonlinear dynamical systems", *Math. Comput. Modelling* **25** (1997), 101-113
99. P. Collet and J.P. Eckmann *Instabilities and fronts in extended systems*, Princeton University Press, Princeton 1990
100. P. Collet and J.P. Eckmann, "Space-time behaviour in problems of hydrodynamic type: a case study", *Nonlinearity* **5** (1992), 1265-1302
101. P.A. Clarkson and M.D. Kruskal: "New similarity reduction of the Boussinesq equation", *J. Math. Phys.* **30** (1989), 2201
102. R. Courant and D. Hilbert, *Methods of Mathematical Physics*, Wiley, New York 1962, 1989
103. W. Craig and C.E. Wayne, "Newton's method and periodic solutions of nonlinear wave equations", *Comm. Pure Appl. Math.* **46** (1993), 1419-1501
104. J.D. Crawford, "Introduction to bifurcation theory", *Rev. Mod. Phys.* **63** (1991), 991-1037
105. J.D. Crawford and E. Knobloch, "Symmetry and symmetry-breaking bifurcations in fluid dynamics", *Ann. Rev. Fluid Mech.* **23** (1991), 342
106. M.C. Cross and P.C. Hohenberg, "Pattern formation outside of equilibrium", *Rev. Mod. Phys.* **65** (1993), 851-1112

107. R. Cushman and J.A. Sanders. "Nilpotent normal forms and representation theory of $sl(2,R)$", in: "Multi-parameter bifurcation theory" (M. Golubitsky and J. Guckenheimer ed.s), *Contemp. Math.* **56**, A.M.S., Providence, 1986
108. A.M.O. de Almeida and W.M. Vieira, "Extended convergence of normal forms around unstable equilibria", *Phys. Letters A* **227** (1997), 298-300
109. A. Degasperis and G. Gaeta eds., *SPT98. Symmetry and Perturbation Theory II* (proceedings Roma workshop 1998), World Scientific, Singapore 1999
110. A. Degasperis, S.V. Manakov and P.M. Santini, "Multiple-scales perturbation beyond the nonlinear Schrödinger equation. I.", *Physica D* **100** (1997), 187-211
111. A. Delshams and P. Gutierrez, "Effective stability and KAM theory", *J. Diff. Eq.* **128** (1996), 415-490
112. A. Delshams and P. Gutierrez, "Estimates on invariant tori near an elliptic equilibrium point of a Hamiltonian system", *J. Diff. Eq.* **131** (1996), 277-303
113. G. Deninno and E. Todesco, "Effect of linear couplings on nonlinear resonances in betatron motion", *Phys. Rev. A* **55** (1997), 2059-2062
114. V.A. Dorodnitsyn, "Transformation groups in jet spaces", *J. Sov. Math.* **55** (1991), 1490
115. A.J. Dragt, "Summary of the working group on maps", *part. Accel.* **55** (1996), 499-530
116. B.A. Dubrovin, I.M. Krichever and S.P. Novikov, *Integrable systems*, in Encyclopaedia of Mathematical Sciences vol. 4 – Dynamical Systems IV (V.I. Arnold and S.P. Novikov eds.), Springer, Berlin 1988
117. B. Dubrovin, S. Novikov and A. Fomenko, *Modern Geometry I, II & III*, Springer, Berlin 1984 & 1989; *Geometrie Contemporaine I, II & III*, Mir, Moscow, 1982 & 1987
118. J.J. Duistermaat, "Non-integrability of the 1:1:2 resonance", *Ergod. Th. & Dyn. Syst.* **4** (1984), 553-568
119. J. Ecalle, "Nongeometric models of singularities", *Ann. Inst. Fourier* **42** (1992), 73-164
120. C. Ehresmann, *C. R. Acad. Sci. Paris* **233** (1951), 598, 777, 1081; **234** (1952), 1028, 1424, 587; **239** (1954), 1762; **240** (1955), 397, 1755. See also: "Les prolongements d'une varieté differentiable", *Atti del IV congresso dell'Unione Matematica Italiana* (1951)
121. C. Elphick, E. Tirapegui, M.E. Brachet, P. Coullet P., and G. Iooss, "A simple global characterization for normal forms of singular vector fields", *Physica D* **29** (1987), 95-127; addendum *Physica D* **32** (1988), 488
122. U.M. Engel, B. Stegemerten, P. Eckelt, "Normal forms and quasi-integrals for the Hamiltonians of magnetic bottles", *J. Phys. A* **28** (1995), 1425-1448
123. F. Fassò, "Lie series method for vector fields and Hamiltonian perturbation theory", *ZAMP* **41** (1990), 843-864
124. F. Fassò, M. Guzzo and G. Benettin, "Nekhoroshev stability of elliptic equilibria of Hamiltonian systems", *Comm. Math. Phys.* **197** (1998), 347-360
125. M.J. Field, "Equivariant bifurcation theory and symmetry breaking", *J. Dyn. Dif. Eqs.* **1** (1989), 369-421
126. M.J. Field, "Symmetry breaking for compact Lie groups", *Mem. A.M.S.* **574** (1996), 1
127. M.J. Field, *Lectures on bifurcations, dynamics and symmetry*, Res. Notes Math. 356, Pitman, Boston 1996

128. M.J. Field and R.W. Richardson, "Symmetry breaking and the maximal isotropy subgroup conjecture for reflection groups", *Arch. Rat. Mech. Anal.* **105** (1989), 61-94
129. M.J. Field and R.W. Richardson, "Symmetry breaking in equivariant bifurcation problems", *Bull. Am. Math. Soc.* **22** (1990), 79-84
130. M.J. Field and R.W. Richardson, "Symmetry breaking and branching patterns in equivariant bifurcation theory I", *Arch. Rat. Mech. Anal.* **118** (1992), 297-348
131. M.J. Field and R.W. Richardson, "Symmetry breaking and branching patterns in equivariant bifurcation theory II", *Arch. Rat. Mech. Anal.* **120** (1992), 147-190
132. A.S. Fokas, "Group theoretical aspects of constants of motion and separable solutions in classical mechanics", *J. Math. Anal. Appl.* **68** (1979), 347-370
133. A.S. Fokas, "Generalized symmetries and constants of motion of evolution equations", *Lett. Math. Phys.* **3** (1979), 467-473
134. A.S. Fokas, "A symmetry approach to exactly solvable evolution equations", *J. Math. Phys.* **21** (1980), 1318-1326
135. A.S. Fokas, "Symmetries and integrability", *Stud. Appl. Math.* **77** (1987), 253-299
136. A.S. Fokas and I.M. Gelfand, "Surfaces on Lie groups, Lie algebras, and their integrability", *Comm. Math. Phys.* **177** (1996), 203-220
137. E. Fontich and V.G. Gelfreich, "On analytical properties of normal forms", *Nonlinearity* **10** (1997), 467-477
138. E. Forest and D. Murray, "Freedom in minimal normal forms", *Physica D* **74** (1994), 181-196
139. J.P. Francoise, "Singularités de champs isochores", *Duke Math. J.* **47** (1980), 465-485
140. W.I. Fushchich, "Conditional symmetries of the equations of mathematical physics", in *Modern group analysis*, N.H. Ibragimov, M. Torrisi and A. Valenti eds., Kluwer, Dordrecht 1993
141. W.I. Fushchich and A.G. Nikitin, *Symmetries of Maxwell equations*, Reidel, Dordrecht 1987
142. W.I. Fushchich, W.M. Shtelen and S.L. Slavutsky, *Symmetry analysis and exact solutions of nonlinear equations of mathematical physics*, Naukova Dumka, Kiev 1989
143. G. Gaeta, "Bifurcation and symmetry breaking", *Phys. Rep.* **189** (1990), 1-87
144. G. Gaeta, "Autonomous systems, dynamical systems, LPTI symmetries, topology of trajectories, and periodic solutions", *Int. J. Theor. Phys.* **32** (1993), 191-199
145. G. Gaeta, "Normal forms for reversible dynamical systems", *Int. J. Theor. Phys.* **33** (1994), 1917-1928
146. G. Gaeta, *Nonlinear symmetries and nonlinear equations*, Kluwer, Dordrecht 1994
147. G. Gaeta, "Asymptotic symmetries and asymptotically symmetric solutions of partial differential equations", *J. Phys. A* **27** (1994), 437-451
148. G. Gaeta, "Reduction of Poincaré normal forms", *Lett. Math. Phys.* **42** (1997), 103
149. G. Gaeta, "Poincaré renormalized forms", *Ann. I.H.P. (Phys. Theo.)* **70** (1999), 461-514
150. G.Gaeta and G.Marmo, "Nonperturbative linearization of dynamical systems", *J. Phys. A* **29** (1996), 5035-5048

151. G. Gaeta and M.A. Rodriguez, "Discrete symmetries of differential equations", *J. Phys. A* **29** (1996), 859-880
152. G. Gallavotti, "Perturbation theory", in *Mathematical Physics towards the XXI century* (R. Sen and A. Gersten eds.), Ben Gurion University Press 1994
153. G. Gallavotti, "Twistless KAM tori, quasi flat homoclinic intersections, and other cancellations in the perturbation series of certain completely integrable Hamiltonian systems: a review", *Rep. Math. Phys.* **6** (1994), 344-411
154. G. Gallavotti, "Lindstedt series and Kolmogorov theorem", *mp-arc 95-152* and *chao-dyn 9503007*
155. Th. Gallay, "A center-stable manifold theorem for differential equations in Banach spaces", *Comm. Math. Phys.* **152** (1993), 249-268
156. G. Gentile and V. Mastropietro, "Tree expansion and multiscale analysis for KAM tori", *Nonlinearity* **8** (1995), 1159-1178
157. G. Gentile and V. Mastropietro, "KAM theorem revisited", *Physica D* **90** (1996), 225-234
158. G. Gentile and V. Mastropietro, "Methods for the analysis of the Lindestedt series for KAM theory and renormalizability in classical mechanics", *Rev. Math. Phys.* **8** (1996), 393-444
159. A. Giorgilli, "Rigorous results on the power expansions for the integrals of a hamiltonian system near an elliptic equilibrium point", *Ann. I.H.P. (Phys. Theo.)* **48** (1988), 423-439
160. A. Giorgilli, A. Delshams, E. Fontich, L. Galgani and C. Simó, "Effective stability for Hamiltonian systems near an elliptic equilibrium point, with an application to the restricted three body problem", *J. Diff. Eqs.* **77** (1989), 167-198
161. A. Giorgilli and U. Locatelli, "Kolmogorov theorem and classical perturbation theory", *ZAMP* **48** (1997), 220-261
162. A. Giorgilli and A. Morbidelli, "Invariant KAM tori and global stability for hamiltonian systems", *ZAMP* **48** (1997), 102-134
163. A. Giorgilli and E. Zehnder, "Exponential stability for time dependent potentials", *ZAMP* **43** (1992), 827-855
164. P. Glendinning, *Stability, instability and chaos: an introduction to the theory of nonlinear differential equations*, Cambridge University Press, Cambridge 1994
165. N. Goldenfeld, *Lectures on phase transitions and the renormalization group*, Addison-Wesley, Reading 1992
166. N. Goldenfeld, O. Martin and Y. Oono, "Intermediate asymptotics and renormalization group theory", *J, Sci. Comp.* **4** (1989), 355-372
167. N. Goldenfeld, O. Martin, Y. Oono and F. Liu, "Anomalous dimensions and the renormalization group in a non-linear diffusion process", *Phys. Rev. Lett.* **65** (1990), 1361-1364
168. M. Golubitsky, J. Marsden and D. Schaeffer, "Bifurcation problems with hidden symmetry", in *Partial differential equations and dynamical systems*, W.E. Fitzgibbon ed., Research Notes in Mathematics **101**, Pitman, Boston 1984
169. M. Golubitsky, I. Stewart and D. Schaeffer, *Singularity and groups in bifurcation theory - vol. II*, Springer, Berlin 1988
170. R.R. Gompa, "Perturbation theories using normal forms", *Nuovo Cim. B* **108** (1993), 967-986
171. A. Gonzalez, "Improved semiclassics for the excited states of three particles", *Few-Body Systems* **20** (1996), 155-173

172. T. Gramtchev and M. Yoshino, "Rapidly convergent iteration methods for simultaneous normal forms of commuting maps", preprint 1998, to appear in *Math. Z.*
173. J. Guckenheimer and P. Holmes, *Nonlinear oscillations, dynamical systems, and bifurcation of vector fields*, Springer, Berlin 1983
174. F.G. Gustavson, "On constructing formal integrals of a Hamiltonian system near an equilibrium point" *Astron. J.* **71** (1964), 670-686
175. M. Guzzo, F. Fassò and G. Benettin, "On the stability of elliptic equilibria", *Math. Phys. El. J.* **4** (1998)
176. M. Guzzo and A. Morbidelli, "Construction of a Nekhoroshev like result for the asteroid belt dynamical system", *Cel. Mech.* **66** (1997), 255-292
177. G.A. Hagedorn, "Classification and normal forms for avoided crossings of quantum-mechanical energy levels", *J. Phys. A* **31** (1998), 369-383
178. I. Halperin, "Quantum KAM technique and Yang-Mills quantum mechanics", *A,, Phys.* **244** (1995), 445-475
179. M. Hamermesh, *Group theory*, Addison-Wesley, Reading 1962; reprinted by Dover, New York 1991
180. W. Hereman, "Review of symbolic software for symmetry analysis of partial differential equations", *Euromath. Bull.* **2** (1994), 45; "Symbolic software for Lie symmetry analysis"; in *CRC Handbook of Lie groups analysis of differential equations – Vol. 3* (N.H. Ibragimov ed.), pp. 365-414, CRC, Boca Raton 1996
181. R. Hermann, "The formal linearization of a semisimple Lie algebra of vector fields about a singular point", *Trans. A.M.S.* **130** (1968), 105-109
182. R. Hermann, *Lie groups for physicists*, Benjamin, New York 1966
183. M.W. Hirsch and S. Smale, *Differential equations, dynamical systems, and linear algebra*, Academic Press, London 1974
184. G.H. Hoffstatter and M. Berz, "Rigorous lower bounds on the survival time in particle accelerators", *Particle Accelerators* **55** (1996), 193-202
185. P. Holmes, "Unfolding a degenerate nonlinear oscillator: a codimension two bifurcation", *Ann. New York Acad. of Sciences* **357** (1980), 473-487
186. I. Hoveijn, "Versal deformations and normal forms for reversible and Hamiltonian linear systems" , *J. Diff. Eq.*, **126** (1996), 408-442
187. I. Hoveijn and F. Verhulst, "Chaos in the 1:2:3 hamiltonian normal form", *Physica D* **44** (1990), 397-406
188. N. Ibragimov, "Group analysis of ordinary differential equations and the invariance principle in Mathematical Physics", *Russ. Math. Surv.* **47** (1992), 89
189. Yu.S. Il'yashenko and S.Yu. Yakovenko, "Finitely smooth normal forms of local families of diffeomorphisms and vector fields", *Russ. Math. Surv.* **46** (1991), 1-43
190. G. Iooss and M. Adelmeyer, "Topics in bifurcation theory and applications, World Scientific, Singapore 1992
191. G. Iooss and D.D. Joseph, *Elementary stability and bifurcation theory*; Springer, Berlin 1990
192. C. Isham, *Modern differential geometry for physicists*, World Scientific, Singapore 1989
193. H. Ito, "Convergence of Birkhoff normal form for integrable systems", *Comm. Math. Helv.* **64** (1989), 412-461
194. H. Ito, "Integrability of Hamiltonian systems and Birkhoff normal forms in the simple resonance case", *Math. Ann.* **292** (1992), 411-444
195. H. Ito, "Integrable symplectic maps and their Birkhoff normal forms", *Tohoku Math. J.* **49** (1997), 73-114

196. G. Jang, V. Vittal and W. Kliemann, "Effect of nonlinear modal interaction on control performance: Use of normal forms technique in control design - Part I: General theory and procedure", *IEEE Trans. Power Syst.* **13** (1998), 401-407
197. G. Jang, V. Vittal and W. Kliemann, "Effect of nonlinear modal interaction on control performance: Use of normal forms technique in control design - Part II: Case studies", *IEEE Trans. Power Syst.* **13** (1998), 408-413
198. A. Jorba and J. Villanueva, "Numerical computation of normal forms around some periodic orbits of the restricted three-body problem", *Physica D* **114** (1998), 197-229
199. M. Kaluza, "Program for Birkhoff-Gustavson normal form for N-degrees of freedom", *Computer Phys. Comm.* **74** (1993), 441-449
200. E. Kalnins, W. Miller and P. Winternitz, "The group O(4), separation of variables and the hydrogen atom", *SIAM J. Appl. Math.* **30** (1976), 630-664
201. T. Kappeler and M. Makarov, "On the symplectic foliation induced by the second Poisson bracket for KdV", in *Symmetry and perturbation theory* (D. Bambusi and G. Gaeta eds.), C.N.R., Roma 1998
202. M. Kinyon and S. Walcher: *On ordinary differential equations admitting a finite linear group of symmetries.* J. Math. Analysis Appl. **216**, 180-196 (1997).
203. A.A. Kirillov, *Elements of the Theory of Representations*; Springer, Berlin 1976, 1984
204. Y. Kodama, "Normal forms, symmetry and infinite dimensional Lie algebra for systems of ODE's", *Phys. Lett. A* **191** (1994), 223-228
205. Y. Kodama and A.V. Mikhailov, "Obstacles to asymptotic integrability", in *Algebraic aspects of integrable systems*, A.S. Fokas and I.M. Gelfand eds., Birkhauser, Boston 1997
206. H. Kokubu, H. Oka H and D. Wang, "Linear grading function and further reduction of normal forms", *J. Diff. Eq* **132** (1996), 293-318
207. V.V. Kozlov, *Symmetries, topology and resonances in Hamiltonian mechanics*, Springer, Berlin 1996
208. S. Kuksin, "Nearly integrable infinite-dimensional hamiltonian systems", *Lect. Notes Math.* **1556**, Springer, Berlin 1994
209. M. Kummer, "How to avoid secular terms in classical and quantum mechanics", *Nuovo Cimento B*, **1** (1971), 123
210. M. Kummer, "On resonant nonlinearly coupled oscillators with two equal frequencies", *Comm. Math. Phys.* **48** (1976), 53
211. J. Lamb, "Local bifurcations in k-symmetric dynamical systems", *Nonlinearity* **9** (1996), 537-557
212. J. Lamb, "k-symmetry and return maps of spacetime symmetric flows", *Nonlinearity* **11** (1998), 601-630
213. J. Lamb and J. Roberts, "Time reversal symmetry in dynamical systems: a survey", *Physica D* **112** (1998), 1-39
214. L.D. Landau and E.M. Lifschitz, *Statistical Physics*; Pergamon, London, 1955
215. L.D. Landau and E.M. Lifschitz, *Quantum Mechanics*; Pergamon, London, 1955
216. Leboeuf Mouchet, "Normal forms and complex periodic orbits in semiclassical expansions of Hamiltonian systems", preprint *chao-dyn 9809015*
217. T.K. Leen, "A coordinate-independent center manifold reduction", *Phys. Lett. A* **174** (1993), 89-93
218. D. Levi and M.A. Rodriguez, "Symmetry groups of partial differential equations and of differential-difference equations: the Toda lattice versus the Korteweg-de Vries equation", *J. Phys. A* **25** (1992), L975

219. D. Levi and P. Winternitz, "Non-classical symmetry reduction: example of the Boussinesq equation", *J. Phys. A* **22** (1989) 2915
220. D. Levi and P. Winternitz, "Continuous symmetries of discrete equations", *Phys. Lett. A* **152** (1991), 336;
221. D. Levi and P. Winternitz, "Symmetries and conditional symmetries of differential-difference equations", *J. Math. Phys.* **34** (1993), 3713
222. D. Levi and P. Winternitz, "Symmetries of discrete dynamical systems", *J. Math. Phys.* **37** (1996), 676-684
223. A.J. Lichtenberg and M.A. Lieberman, *Regular and stochastic motion*, Springer, Berlin 1983
224. C.M. Lin, V. Vittal, W. Kliemann and A.A. Fouad, "Investigation of modal interaction and its effects on control performance in stressed power systems using normal forms of vector fields", *IEEE Trans. Power Syst.* **11** (1996), 781-787
225. P. Lochak and C. Meunier, *Multiphase averaging methods for Hamiltonian systems*, Springer, Berlin 1988
226. S. Louis and L. Brenig, "Structure and convergence of Poincaré normal forms", *Phys. Lett A* **233** (1997) 184-192
227. R.S. Mackay and J.D. Meiss eds., *Hamiltonian dynamical systems, a reprint collection*, Adam Hilgher & IOP, Bristol 1987
228. P. Manneville, *Dissipative structures and weak turbulence*; Academic Press, London 1990
229. L.M. Markhashov, "On the reduction of differential equations to the normal form by an analytic transformation" *J. Appl. Math. Mech.* **38** (1974), 788-790
230. J.E. Marsden, *Lectures on Mechanics*, Cambridge University Press, Cambridge 1992
231. J.E. Marsden and M. Mc Cracken (eds.), *The Hopf Bifurcation and its Applications*, Springer, Berlin 1976
232. J.E. Marsden and T. Ratiu, *Introduction to mechanics and symmetry*, Springer, Berlin 1994
233. M. Mazzocco, "KAM theorem for generic analytic perturbations of the Euler system", *ZAMP* **48** (1997), 193-219
234. S.M. Mcrae and E.R. Vrscay, "Perturbation limit and the classical limit of quantum mechanics", *J. Math. Phys.* **38** (1997), 2899-2921
235. I. Melbourne and M. Dellnitz, "Normal forms for linear Hamiltonian vector-fields commuting with the action of a compact Lie group", *Math. Proc. Cambridge Phil. Soc.* **114** (1993), 235-268
236. L. Michel, "Symmetry defects and broken symmetry. Configurations. Hidden symmetry", *Rev. Mod. Phys.* **52** (1980), 617
237. A.V. Mikhailov, A.B. Shabat and R.I. Yamilov, "The symmetry approach to the classification of non-linear equations. Complete list of integrable systems", *Russ. Math. Surv.* **42** (1987), 1
238. W. Miller, *Symmetry groups and their applications*; Academic Press, London 1972
239. W. Miller, *Symmetry and separation of variables*, Addison-Wesley, Reading 1977
240. Yu. A. Mitropolsky and A.K. Lopatin, "Nonlinear mechanics, groups and symmetry", Kluwer, Dordrecht 1995
241. J. Montaldi, M. Roberts and I. Stewart, "Periodic solutions near equilibria of symmetric Hamiltonian systems", *Phil. Trans. Roy. Soc. London A* **326** (1988), 237-293

242. J. Montaldi, M. Roberts and I. Stewart, "Existence of nonlinear normal modes in symmetric Hamiltonian systems", *Nonlinearity* **3** (1990), 695-730.
243. J. Montaldi, M. Roberts and I. Stewart, "Stability of nonlinear normal modes in symmetric Hamiltonian systems", *Nonlinearity* **3** (1990), 731-772
244. A. Morbidelli and A. Giorgilli, "Superexponential stability of KAM theory", *J. Stat. Phys.* **78** (1995), 1607-1617
245. A. Morbidelli and A. Giorgilli, "On the role of high order resonances in normal forms and in separatrix splitting", *Physica D* **102** (1997), 195-207
246. J.K. Moser: "New aspects in the theory of stability of Hamiltonian systems", *Comm. Pure Appl. Math.* **11** (1958), 81-114
247. J.K. Moser, "Lectures on Hamiltonian systems" *Mem. Amer. Math. Soc.* **81** (1968), 1-60
248. J.K. Moser, "Regularization of Kepler's problem and the averaging method on a manifold", *Comm. Pure Appl. Math.* **23** (1970), 609-636
249. G.D. Mostow, "Equivariant embeddings in euclidean space", *Ann. Math.* **65** (1957), 432-446
250. M.A. Naimark and A.I. Stern, *Theory of Group Representations*, Springer, Berlin 1982
251. A.H. Nayfeh, *Method of normal forms*, Wiley, New York 1993
252. A.I. Neishtadt, "Averaging in multifrequency systems", *Sov. Phys. Dokl.* **20** (1975), 492-494
253. A.I. Neishtadt, "Averaging in multifrequency systems II", *Sov. Phys. Dokl.* **21** (1976), 80-82
254. N.N. Nekhoroshev, "Behaviour of hamiltonian systems close to integrable", *Funct. Anal. Appl.* **5** (1971), 339-340
255. N.N. Nekhoroshev, "An exponential estimate of the time of stability of nearly integrable Hamiltonian systems", *Russ. Math. Surv.* **32** (1977), 1-65
256. L. Niedermann, "Stability over exponential long times in the planetary problem", *Nonlinearity* **9** (1996), 1703-1751
257. L. Niedermann, "Nonlinear stability around an elliptic equilibrium point in a Hamiltonian system", *Nonlinearity* **11** (1998), 1465-1479
258. A.S. Nikolaev, "On the diagonalization of quantum Birkhoff-Gustavson normal form", *J. Math. Phys.* **37** (1996), 2643-2661
259. N.V. Nikolenko, "The Method of Poincaré Normal Form in Problems of Integrability of Equations of Evolution Type", *Russ. Math. Surv.* **41** (1986), 63-114
260. M.C. Nucci, "The role of symmetries in solving differential equations", *Math. Comp. Modelling* **25** (1997), 181-193
261. M.A. Olshanetsky and A.M. Perelomov, *Integrable systems and finite dimensional Lie algebra*, in Encyclopaedia of Mathematical Sciences vol. 16 – Dynamical Systems VII, (V.I. Arnold and S.P. Novikov eds.), pp. 87-115; Springer, Berlin 1994
262. P.J. Olver, *Applications of Lie groups to differential equations*; Springer, Berlin 1986
263. P.J. Olver, *Equivalence, Invariants, and Symmetry*; Cambridge University Press, Cambridge 1995
264. P.J. Olver and P. Rosenau, "Group-invariant solutions of differential equations", *SIAM J. Appl. Math.* **47** (1987), 263-278
265. L.V. Ovsjiannikov, *Group properties of differential equations*; Novosibirsk, 1962 (English transl. by G.W. Bluman, 1967)

266. L.V. Ovsjiannikov, *Group analysis of differential equations*; Academic Press, London 1982
267. J. Pade, A. Rauh, G. Tsarouhas, "Application of normal forms to the Lorenz model in the subcritical region" *Physica D* **29** (1987), 236-246
268. R. Palais, "Imbedding of compact, differentiable transformation groups in orthogonal representations", *J. Math. Mech.* **6** (1957), 673-678
269. A.M. Perelomov, *Integrable systems in classical Mechanics and Lie algebras*, Birkhauser, Boston 1990
270. H. Poincaré, *Analysis situ*, reprinted in: "Oeuvres Completes, vol. VI", Gabay, Paris 1996
271. J. Pöschel, "Nekhoroshev estimates for quasi-convex Hamiltonian systems", *Math. Z.* **213** (1993), 187-216
272. E. Pucci and G. Saccomandi, "On the weak symmetry group of partial differential equations", *J. Math. Anal. Appl.* **163** (1992), 588-598
273. E. Pucci and G. Saccomandi, "Potential symmetries and solutions by reduction of partial differential equations", *J. Phys. A* **26** (1993), 681-690
274. E. Pucci, "Similarity solutions of PDEs", *J. Phys. A* **25** (1992), 2631
275. L.D. Pustylnikov, "Infinite dimensional nonlinear ordinary differential equations and the KAM theory", *Russ. Math. Surv.* **52** (1997), 551-604
276. R.H. Rand and D. Armbruster, "Perturbation methods, bifurcation theory and computer algebra", Springer, Berlin 1987
277. A.G. Reyman and M.A. Semenov Tian Shansky, *Group theoretical methods in the theory of finite dimensional integrable systems*, in Encyclopaedia of Mathematical Sciences vol. 16 – Dynamical Systems VII, (V.I. Arnold and S.P. Novikov eds.), pp. 116-225; Springer, Berlin 1994
278. D. Ruelle, "Bifurcation in the presence of a symmetry group", *Arch. Rat. Mech. Anal.* **51** (1973), 136
279. D. Ruelle, *Elements of Differentiable Dynamics and Bifurcation Theory*, Academic Press, London 1989
280. H. Rüssmann, "Über das Verhalten analytischer Hamiltonscher Differentialgleichungen in der Nähe einer Gleichgewichtslösung" *Math. Ann.* **154** (1964), 285-300
281. D.A. Sadovskii and J.B. Delos, "Bifurcation of the periodic orbits of Hamiltonian systems – an analysis using normal form theory", *Phys. Rev. A* **54** (1996) 2033-2070
282. D.A. Sadovskii, J.A. Shaw and J.B. Delos, "Organization of sequences of bifurcations of periodic orbits", *Phys. Rev. Lett.* **75** (1995), 2120-2123
283. D.A. Sadovskii and B.I. Zhilinskii, "Tuning the hydrogen atom in crossed fields between the Zeeman and Stark limits", *Phys. Rev. A* **57** (1998), 2867-2884
284. A.W. Saenz, "Long-time approximation to the evolution of resonant and nonresonant anharmonic oscillators in quantum mechanics", *J. Math. Phys.* **37** (1996) 2182-2205
285. J.A. Sanders and F. Verhulst, *Averaging methods in nonlinear dynamical systems*, Springer, Berlin 1985
286. W. Sarlet, P.G.L. Leach and F. Cantrijn, "First integrals versus configurational invariants and a weak form of complete integrability", *Physica D* **17** (1985), 87-98
287. D.H. Sattinger, *Topics in stability and bifurcation theory*; Lecture Notes in Mathematics **309**, Springer, Berlin 1973

288. D.H. Sattinger, *Group theoretic methods in bifurcation theory*; Lecture Notes in Mathematics **762**, Springer, Berlin 1979
289. D.H. Sattinger, *Branching in the presence of symmetry*; SIAM, Philadelphia 1983
290. D.H. Sattinger and O. Weaver, *Lie groups and algebras*; Springer, Berlin 1986
291. D.J. Saunders, *The geometry of jet bundles*; Cambridge University Press, Cambridge 1991
292. B. Schuman, "On the Birkhoff normal form and isochronous centre", *Comptes Rendus Acad. Sci. Serie I-Math.* **323** (1996), 21-24
293. C.A. Schwartz and A.G. Yan, "Construction of Lyapunov functions for nonlinear systems using normal forms", *J. Math. Anal. Appl.* **216** (1997), 521-535
294. M.B. Sevryuk, "Some problems in the KAM theory. Conditionally-periodic motions in typical systems", *Russ. Math. Surv.* **50** (1995), 342-353
295. M.B. Sevryuk, "Reversible systems", *Lect. Notes Math.* **1211**, Springer, Berlin 1986
296. I.I. Shevchenko and A.G. Sokolsky, "Hyperboloidal precession of a dynamically symmetric satellite – construction of normal forms of the Hamiltonian", *Cel. Mech.* **62** (1995), 289-304
297. A.N.Shoshitaishvili, "Bifurcations of topological type at singular points of parametrized vector fields" *Funct. Anal. Appl.* **6** (1972), 169
298. A.N.Shoshitaishvili, "The bifurcation of the topological type of the singular points of vector fields that depend on parameters" *Tr. Semin. I.G. Petrovskogo* **1** (1975), 279
299. K. Siegel and J.K. Moser, *Lectures on Celestial Mechanics*, Springer, Berlin 1971, reprinted in "Classics in Mathematics" 1995
300. S. Smale, *The mathematics of time: essays on dynamical systems, economic processes and related topics*, Springer, Berlin 1990
301. V.V. Sokolov, "On the symmetries of evolutions equations", *Russ. Mah. Surv.* **43** (1988), 165
302. V.V. Sokolov and A.B. Shabat, "Classification of integrable evolution equations", in *Mathematical Physics Reviews – vol.4*, S.P. Novikov ed., Soviet Scientific Reviews, 1984
303. K. Sohlberg and R.B. Shirts, "Symmetry of approximate Hamiltonians generated in Birkhoff-Gustavson normal form", *Phys. Rev. A* **54** (1996), 416-422
304. M. Spivak, *A comprehensive introduction to Differential Geometry*; Publish or Perish, Berkeley 1970, 1979
305. H. Stephani, *Differential equations. Their solution using symmetries*, Cambridge University Press, Cambridge 1989
306. S. Sternberg, "Local contractions and a theorem of Poincaré", *Amer. J. Math.* **79** (1957), 809-824
307. S. Sternberg, "On the structure of local homeomorphisms of Euclidean n-spaces", *Amer. J. Math.* **80** (1958), 623-631
308. I. Stewart, "Bifurcation with symmetry", in *New directions in dynamical systems*, T. Bedford and J. Swift eds., Cambridge University Press, Cambridge 1988
309. F. Takens, "Partially hyperbolic fixed points" *Topology* **10** (1971), 133
310. K.E. Thylwe and H. Dankowicz, "Time-dependent normal form Hamiltonian for dynamical equilibria", *J. Phys. A: Math. Gen.* **29** (1996), 3707-3722
311. E. Todesco, "Local analysis of formal stability and existence of fixed points in 4D symplectic mappings", *Physica D* **95** (1996), 1-12

312. V.V. Trofimov and A.T. Fomenko, *Geometric and algebraic mechanism of the integrability of hamiltonian systems on homogeneous spaces and Lie algebras*, in Encyclopaedia of Mathematical Sciences vol. 16 – Dynamical Systems VII, (V.I. Arnold and S.P. Novikov eds.), pp. 87-115; Springer, Berlin 1994
313. G. Turchetti, "Geometry of orbits and action diffusion from normal forms in nonlinear beam dynamics", *Part. Accel.* **55** (1996), 167-181
314. A. Vanderbauwhede, *Local bifurcation and symmetry*; Pitman, Boston 1982
315. A. Vanderbauwhede, "Center manifolds, normal forms, and elementary bifurcations"; *Dynamics Reported* **2** (1989), 89-169
316. A. Vanderbauwhede and G. Iooss, "Centre manifold theory in infinite dimensions", *Dynamics Reported (new series)* **1** (1992), 125-163
317. J-C. van der Meer, *The Hamiltonian Hopf bifurcation*; Lecture Notes in Mathematics **1160**, Springer, Berlin 1985
318. G. Vattay, "Bohr-Sommerfeld quantization of periodic orbits", *Phys. Rev. Lett.* **76** (1996), 1059-1062
319. F. Verhulst, *Nonlinear differential equations and dynamical systems*, Springer, Berlin 1989, 1996
320. F. Verhulst, "Symmetry and integrability in Hamiltonian normal form", in *Symmetry and perturbation theory* (D. Bambusi and G. Gaeta eds.), C.N.R., Roma 1998
321. F. Verhulst, "On averaging methods for partial differential equations", in *Symmetry and perturbation theory II* (A. Degasperis and G. Gaeta eds.), World Scientific, Singapore 1999
322. F. Verhulst and I. Hoveijn, "Chaos in the 1:2:3 Hamiltonian normal forms", *Physica D* **44** (1990), 397-406
323. F. Verhulst and I. Hoveijn, "Integrability and chaos in hamiltonian normal forms", in *Geometry and analysis in nonlinear dynamics* (H.W. Broer and F. Takens eds.), Pitman, Boston 1992
324. J. Vey, "Sur certain systèmes dynamiques séparables" *Amer. J. Math.* **100** (1978), 591-614
325. W. Vieira and A.M.O. de Almeida, "Study of chaos in Hamiltonian systems via convergent normal forms", *Physica D* **90** (1996), 9-30
326. A.M. Vinogradov, "Local symmetries and conservation laws", *Acta Appl. Math.* **2** (1984), 21
327. V. Vittal, W. Kliemann, Y.X. Ni, D.G. Chapman, A.D. Silk and D.J. Sobajic, "Determination of generator groupings for an islanding scheme in the Manitoba Hydro system using the method of normal forms" *IEEE Trans power syst.* **13** (1998), 1346-1351
328. E.M. Vorob'ev, "Partial symmetries of systems of differential equations", *Soviet Math. Dokl.* **33** (1986), 408
329. E.M. Vorob'ev, "Reduction and quotient equations for differential equations with symmetries", *Acta Appl. Math.* **23** (1991), 1
330. S. Walcher, "On differential equations in normal form", *Math. Ann.* **291** (1991), 293-314
331. S. Walcher, "On transformation into normal form", *J. Math. Anal. Appl.* **180** (1993), 617-632
332. S. Walcher, "Symmetries of ordinary differential equations", *Nova J. Alg. Geom.* **3** (1993), 245-275

333. S. Walcher, "Orbital symmetries of first order ODEs", in *Symmetry and perturbation theory II* (A. Degasperis and G. Gaeta eds.), World Scientific, Singapore 1999
334. S. Walcher, "On convergent normal form transformations in the presence of symmetry", Preprint 1999, to appear in *J. Math. Anal. Appl.*
335. L. Wang, D.L. Bosley and J. Kevorkian, "An asymptotic analysis of the 1/3/4 hamiltonian normal form", *Physica D* **99** (1996), 1-17
336. W.H. Warner, P.R. Sethna and J.P. Sethna, "A generalization of the theory of normal forms", *J. Nonlinear Science* **6** (1996), 499-506
337. C.E. Wayne, "Periodic and quasi-periodic solutions of nonlinear wave equations via KAM theory", *Comm. Math. Phys.* **127** (1990), 479-528
338. A. Weinstein, "Normal modes for nonlinear Hamiltonian systems" *Inventiones Math.*, **20** (1973) 45-57
339. H. Weyl, *The theory of groups and Quantum Mechanics*, Princeton University Press, Princeton 1950
340. G. Williamson, "On the algebraic problem concerning the normal form of linear dynamical systems", *Am. J. Math.* **58** (1936), 141
341. P. Winternitz, "What is new in the study of differential equations by group theoretical methods?", in *Group Theoretical Methods in Physics* (proceedings of the XV ICGTMP), R. Gilmore ed., World Scientific, Singapore 1987
342. P. Winternitz, "Lie groups and solutions of nonlinear PDEs", in *Integrable systems, quantum groups, and quantum field theory* (NATO ASI 9009), L.A. Ibort and M.A. Rodriguez eds., Kluwer, Dordrecht 1993
343. V.E. Zakharov ed., *What is integrability?*, Springer, Berlin 1991
344. E. Zehnder, "Siegel Linearization Theorem in Infinite Dimension", *Man. Math.* **23** (1978), 363-371
345. W.Y. Zhang and Y.S. Chen, "Adjoint operator method and normal forms of higher order for nonlinear dynamical system", *Appl. Math. Mech. (PMM)* **18** (1997), 449-461
346. W.Y. Zhang, K. Huseyin and Y.S. Chen, "A new approach for obtaining normal forms of non-linear systems", *J. Sound Vibration* **210** (1998), 609-625
347. M. Zhitomirskii, "Normal forms of symmetrical Hamiltonian systems", *Journ. Diff. Eq.* **111** (1994), 58-78

Druck: Strauss Offsetdruck, Mörlenbach
Verarbeitung: Schäffer, Grünstadt

Lecture Notes in Physics

For information about Vols. 1–494
please contact your bookseller or Springer-Verlag

Vol. 495: Y. Kosmann-Schwarzbach, B. Grammaticos, K.M. Tamizhmani (Eds.), Integrability of Nonlinear Systems. VII, 380 pages. 1997.

Vol. 496: F. Lenz, H. Grießhammer, D. Stoll (Eds.), Lectures on QCD. VII, 483 pages. 1997.

Vol. 497: J. P. Greve, R. Blomme, H. Hensberge (Eds.), Stellar Atmospheres: Theory and Observations. VIII, 352 pages. 1997

Vol. 498: Z. Horváth, L. Palla (Eds.), Conformal Field Theories and Integrable Models. Proceedings, 1996. X, 251 pages. 1997.

Vol. 499: K. Jungmann, J. Kowalski, I. Reinhard, F. Träger (Eds.), Atomic Physics Methods in Modern Research, IX, 448 pages. 1997.

Vol. 500: D. Joubert (Ed.), Density Functionals: Theory and Applications, XVI, 194 pages. 1998.

Vol. 501: J. Kertész, I. Kondor (Eds.), Advances in Computer Simulation. VIII, 166 pages. 1998.

Vol. 502: H. Aratyn, T. D. Imbo, W.-Y. Keung, U. Sukhatme (Eds.), Supersymmetry and Integrable Models. Proceedings, 1997. XI, 379 pages. 1998.

Vol. 503: J. Parisi, S. C. Müller, W. Zimmermann (Eds.), A Perspective Look at Nonlinear Media. From Physics to Biology and Social Sciences. VIII, 372 pages. 1998.

Vol. 504: A. Bohm, H.-D. Doebner, P. Kielanowski (Eds.), Irreversibility and Causality. Semigroups and Rigged Hilbert Spaces. XIX, 385 pages. 1998.

Vol. 505: D. Benest, C. Froeschlé (Eds.), Impacts on Earth. XVII, 223 pages. 1998.

Vol. 506: D. Breitschwerdt, M. J. Freyberg, J. Trümper (Eds.), The Local Bubble and Beyond. Proceedings, 1997. XXVIII, 603 pages. 1998.

Vol. 507: J. C. Vial, K. Bocchialini, P. Boumier (Eds.), Space Solar Physics. Proceedings, 1997. XIII, 296 pages. 1998.

Vol. 508: H. Meyer-Ortmanns, A. Klümper (Eds.), Field Theoretical Tools for Polymer and Particle Physics. XVI, 258 pages. 1998.

Vol. 509: J. Wess, V. P. Akulov (Eds.), Supersymmetry and Quantum Field Theory. Proceedings, 1997. XV, 405 pages. 1998.

Vol. 510: J. Navarro, A. Polls (Eds.), Microscopic Quantum Many-Body Theories and Their Applications. Proceedings, 1997. XIII, 379 pages. 1998.

Vol. 511: S. Benkadda, G. M. Zaslavsky (Eds.), Chaos, Kinetics and Nonlinear Dynamics in Fluids and Plasmas. Proceedings, 1997. VIII, 438 pages. 1998.

Vol. 512: H. Gausterer, C. Lang (Eds.), Computing Particle Properties. Proceedings, 1997. VII, 335 pages. 1998.

Vol. 513: A. Bernstein, D. Drechsel, T. Walcher (Eds.), Chiral Dynamics: Theory and Experiment. Proceedings, 1997. IX, 394 pages. 1998.

Vol. 514: F. W. Hehl, C. Kiefer, R. J. K. Metzler, Black Holes: Theory and Observation. Proceedings, 1997. XV, 519 pages. 1998.

Vol. 515: C.-H. Bruneau (Ed.), Sixteenth International Conference on Numerical Methods in Fluid Dynamics. Proceedings. XV, 568 pages. 1998.

Vol. 516: J. Cleymans, H. B. Geyer, F. G. Scholtz (Eds.), Hadrons in Dense Matter and Hadrosynthesis. Proceedings, 1998. XII, 253 pages. 1999.

Vol. 517: Ph. Blanchard, A. Jadczyk (Eds.), Quantum Future. Proceedings, 1997. X, 244 pages. 1999.

Vol. 518: P. G. L. Leach, S. E. Bouquet, J.-L. Rouet, E. Fijalkow (Eds.), Dynamical Systems, Plasmas and Gravitation. Proceedings, 1997. XII, 397 pages. 1999.

Vol. 519: R. Kutner, A. Pękalski, K. Sznajd-Weron (Eds.), Anomalous Diffusion. From Basics to Applications. Proceedings, 1998. XVIII, 378 pages. 1999.

Vol. 520: J. A. van Paradijs, J. A. M. Bleeker (Eds.), X-Ray Spectroscopy in Astrophysics. EADN School X. Proceedings, 1997. XV, 530 pages. 1999.

Vol. 521: L. Mathelitsch, W. Plessas (Eds.), Broken Symmetries. Proceedings, 1998. VII, 299 pages. 1999.

Vol. 522: J. W. Clark, T. Lindenau, M. L. Ristig (Eds.), Scientific Applications of Neural Nets. Proceedings, 1998. XIII, 288 pages. 1999.

Vol. 523: B. Wolf, O. Stahl, A. W. Fullerton (Eds.), Variable and Non-spherical Stellar Winds in Luminous Hot Stars. Proceedings, 1998. XX, 424 pages. 1999.

Vol. 524: J. Wess, E. A. Ivanov (Eds.), Supersymmetries and Quantum Symmetries. Proceedings, 1997. XX, 442 pages. 1999.

Vol. 525: A. Ceresole, C. Kounnas, D. Lüst, S. Theisen (Eds.), Quantum Aspects of Gauge Theories, Supersymmetry and Unification. Proceedings, 1998. X, 511 pages. 1999.

Vol. 526: H.-P. Breuer, F. Petruccione (Eds.), Open Systems and Measurement in Relativistic Quantum Theory. Proceedings, 1998. VIII, 240 pages. 1999.

Vol. 527: D. Reguera, J. M. G. Vilar, J. M. Rubí (Eds.), Statistical Mechanics of Biocomplexity. Proceedings, 1998. XI, 318 pages. 1999.

Vol. 528: I. Peschel, X. Wang, M. Kaulke, K. Hallberg (Eds.), Density-Matrix Renormalization. Proceedings, 1998. XVI, 355 pages. 1999.

Vol. 529: S. Biringen, H. Örs, A. Tezel, J.H. Ferziger (Eds.), Industrial and Environmental Applications of Direct and Large-Eddy Simulation. Proceedings, 1998. XVI, 301 pages. 1999.

Vol. 530: H.-J. Röser, K. Meisenheimer (Eds.), The Radio Galaxy Messier 87. Proceedings, 1997. XIII, 342 pages. 1999.

Vol. 531: H. Benisty, J.-M. Gérard, R. Houdré, J. Rarity, C. Weisbuch (Eds.), Confined Photon Systems. Proceedings, 1998. X, 496 pages. 1999.

Vol. 533: K. Hutter, Y. Wang, H. Beer (Eds.), Advances in Cold-Region Thermal Engineering and Sciences. Proceedings, 1999. XIV, 608 pages. 1999.

Monographs

For information about Vols. 1–15 please contact your bookseller or Springer-Verlag

Vol. m 16: P. Christe, M. Henkel, Introduction to Conformal Invariance and Its Applications to Critical Phenomena. XV, 260 pages. 1993.

Vol. m 17: M. Schoen, Computer Simulation of Condensed Phases in Complex Geometries. X, 136 pages. 1993.

Vol. m 18: H. Carmichael, An Open Systems Approach to Quantum Optics. X, 179 pages. 1993.

Vol. m 19: S. D. Bogan, M. K. Hinders, Interface Effects in Elastic Wave Scattering. XII, 182 pages. 1994.

Vol. m 20: E. Abdalla, M. C. B. Abdalla, D. Dalmazi, A. Zadra, 2D-Gravity in Non-Critical Strings. IX, 319 pages. 1994.

Vol. m 21: G. P. Berman, E. N. Bulgakov, D. D. Holm, Crossover-Time in Quantum Boson and Spin Systems. XI, 268 pages. 1994.

Vol. m 22: M.-O. Hongler, Chaotic and Stochastic Behaviour in Automatic Production Lines. V, 85 pages. 1994.

Vol. m 23: V. S. Viswanath, G. Müller, The Recursion Method. X, 259 pages. 1994.

Vol. m 24: A. Ern, V. Giovangigli, Multicomponent Transport Algorithms. XIV, 427 pages. 1994.

Vol. m 25: A. V. Bogdanov, G. V. Dubrovskiy, M. P. Krutikov, D. V. Kulginov, V. M. Strelchenya, Interaction of Gases with Surfaces. XIV, 132 pages. 1995.

Vol. m 26: M. Dineykhan, G. V. Efimov, G. Ganbold, S. N. Nedelko, Oscillator Representation in Quantum Physics. IX, 279 pages. 1995.

Vol. m 27: J. T. Ottesen, Infinite Dimensional Groups and Algebras in Quantum Physics. IX, 218 pages. 1995.

Vol. m 28: O. Piguet, S. P. Sorella, Algebraic Renormalization. IX, 134 pages. 1995.

Vol. m 29: C. Bendjaballah, Introduction to Photon Communication. VII, 193 pages. 1995.

Vol. m 30: A. J. Greer, W. J. Kossler, Low Magnetic Fields in Anisotropic Superconductors. VII, 161 pages. 1995.

Vol. m 31 (Corr. Second Printing): P. Busch, M. Grabowski, P.J. Lahti, Operational Quantum Physics. XII, 230 pages. 1997.

Vol. m 32: L. de Broglie, Diverses questions de mécanique et de thermodynamique classiques et relativistes. XII, 198 pages. 1995.

Vol. m 33: R. Alkofer, H. Reinhardt, Chiral Quark Dynamics. VIII, 115 pages. 1995.

Vol. m 34: R. Jost, Das Märchen vom Elfenbeinernen Turm. VIII, 286 pages. 1995.

Vol. m 35: E. Elizalde, Ten Physical Applications of Spectral Zeta Functions. XIV, 224 pages. 1995.

Vol. m 36: G. Dunne, Self-Dual Chern-Simons Theories. X, 217 pages. 1995.

Vol. m 37: S. Childress, A.D. Gilbert, Stretch, Twist, Fold: The Fast Dynamo. XI, 406 pages. 1995.

Vol. m 38: J. González, M. A. Martín-Delgado, G. Sierra, A. H. Vozmediano, Quantum Electron Liquids and High-Tc Superconductivity. X, 299 pages. 1995.

Vol. m 39: L. Pittner, Algebraic Foundations of Non-Com-mutative Differential Geometry and Quantum Groups. XII, 469 pages. 1996.

Vol. m 40: H.-J. Borchers, Translation Group and Particle Representations in Quantum Field Theory. VII, 131 pages. 1996.

Vol. m 41: B. K. Chakrabarti, A. Dutta, P. Sen, Quantum Ising Phases and Transitions in Transverse Ising Models. X, 204 pages. 1996.

Vol. m 42: P. Bouwknegt, J. McCarthy, K. Pilch, The W3 Algebra. Modules, Semi-infinite Cohomology and BV Algebras. XI, 204 pages. 1996.

Vol. m 43: M. Schottenloher, A Mathematical Introduction to Conformal Field Theory. VIII, 142 pages. 1997.

Vol. m 44: A. Bach, Indistinguishable Classical Particles. VIII, 157 pages. 1997.

Vol. m 45: M. Ferrari, V. T. Granik, A. Imam, J. C. Nadeau (Eds.), Advances in Doublet Mechanics. XVI, 214 pages. 1997.

Vol. m 46: M. Camenzind, Les noyaux actifs de galaxies. XVIII, 218 pages. 1997.

Vol. m 47: L. M. Zubov, Nonlinear Theory of Dislocations and Disclinations in Elastic Body. VI, 205 pages. 1997.

Vol. m 48: P. Kopietz, Bosonization of Interacting Fermions in Arbitrary Dimensions. XII, 259 pages. 1997.

Vol. m 49: M. Zak, J. B. Zbilut, R. E. Meyers, From Instability to Intelligence. Complexity and Predictability in Nonlinear Dynamics. XIV, 552 pages. 1997.

Vol. m 50: J. Ambjørn, M. Carfora, A. Marzuoli, The Geometry of Dynamical Triangulations. VI, 197 pages. 1997.

Vol. m 51: G. Landi, An Introduction to Noncommutative Spaces and Their Geometries. XI, 200 pages. 1997.

Vol. m 52: M. Hénon, Generating Families in the Restricted Three-Body Problem. XI, 278 pages. 1997.

Vol. m 53: M. Gad-el-Hak, A. Pollard, J.-P. Bonnet (Eds.), Flow Control. Fundamentals and Practices. XII, 527 pages. 1998.

Vol. m 54: Y. Suzuki, K. Varga, Stochastic Variational Approach to Quantum-Mechanical Few-Body Problems. XIV, 324 pages. 1998.

Vol. m 55: F. Busse, S. C. Müller, Evolution of Spontaneous Structures in Dissipative Continuous Systems. X, 559 pages. 1998.

Vol. m 56: R. Haussmann, Self-consistent Quantum Field Theory and Bosonization for Strongly Correlated Electron Systems. VIII, 173 pages. 1999.

Vol. m 57: G. Cicogna, G. Gaeta, Symmetry and Perturbation Theory in Nonlinear Dynamics. XI, 208 pages. 1999.

Vol. m 58: J. Daillant, A. Gibaud (Eds.), X-Ray and Neutron Reflectivity: Principles and Applications. XVIII, 331 pages. 1999.

Vol. m 59: M. Kriele, Spacetime. Foundations of General Relativity and Differential Geometry. XV, 432 pages. 1999.

Batch number: 10682215

Printed by Printforce, the Netherlands